# ENVIRONMENTAL SCIENCE
## MODEL ANSWERS

## Contents

### The Earth's Systems

| | | |
|---|---|---|
| 11 | The Earth's History | 3 |
| 13 | Fossil Formation | 3 |
| 14 | The Earth and the Sun | 3 |
| 15 | The Earth's Crust | 3 |
| 16 | Plate Boundaries | 3 |
| 18 | Lithosphere and Asthenosphere | 3 |
| 19 | Volcanoes and Volcanism | 4 |
| 21 | The Rock Cycle | 4 |
| 22 | Soil Textures | 4 |
| 23 | Soil and Soil Dynamics | 4 |
| 25 | The Atmosphere and Climate | 4 |
| 27 | Variation and Oscillation | 5 |
| 29 | Ocean Circulation and Currents | 5 |
| 31 | Global Water Resources | 5 |
| 33 | Water and People | 6 |
| 35 | Water and Industry | 6 |
| 37 | KEY TERMS: Mix and Match | 7 |

### Ecosystems

| | | |
|---|---|---|
| 40 | Components of an Ecosystem | 7 |
| 41 | Factors Affecting Biome Distribution | 7 |
| 42 | World Distribution of Biomes | 7 |
| 44 | Effect of Temperature on Biomes | 7 |
| 45 | Physical Factors and Gradients | 7 |
| 46 | Physical Factors in a Forest | 7 |
| 47 | Stratification in a Forest | 8 |
| 48 | Physical Factors on a Rocky Shore | 8 |
| 49 | Physical Factors in a Small Lake | 8 |
| 50 | Habitat | 8 |
| 51 | Ecological Niche | 8 |
| 52 | Energy Inputs and Outputs | 9 |
| 53 | Photosynthesis | 9 |
| 54 | Measuring Primary Production | 9 |
| 55 | Cellular Respiration | 9 |
| 56 | Food Chains | 9 |
| 57 | Food Webs | 9 |
| 59 | Energy Flow in an Ecosystem | 10 |
| 61 | Production and Trophic Efficiency | 10 |
| 63 | Ecological Pyramids | 11 |
| 65 | Species Interactions in Communities | 11 |
| 67 | KEY TERMS: Mix and Match | 11 |

### Natural Ecosystem Change

| | | |
|---|---|---|
| 69 | Ecosystem Stability | 11 |
| 71 | Environmental Change | 12 |
| 72 | Nutrient Cycles | 12 |
| 73 | The Carbon Cycle | 12 |
| 75 | The Nitrogen Cycle | 13 |
| 77 | Nitrogen Pollution | 13 |
| 79 | The Hydrological Cycle | 14 |
| 80 | The Phosphorus Cycle | 14 |
| 81 | The Sulfur Cycle | 14 |
| 82 | Primary Succession | 14 |
| 83 | Secondary Succession | 14 |
| 85 | A Case Study in Succession: Surtsey Island | 14 |
| 86 | KEY TERMS: Mix and Match | 15 |

### Populations

| | | |
|---|---|---|
| 88 | The Rise and Fall of Human Populations | 15 |
| 89 | Features of Populations | 15 |
| 90 | Density and Distribution | 15 |
| 91 | Population Regulation | 15 |
| 92 | Population Growth | 16 |
| 93 | Survivorship Curves | 16 |
| 94 | Life Expectancy and Survivorship | 16 |
| 95 | Population Growth Curves | 16 |
| 96 | *r* and *K* Selection | 16 |
| 97 | Population Age Structure | 16 |
| 99 | World Population Growth | 17 |
| 101 | Human Demography | 17 |
| 103 | Human Sustainability | 17 |
| 104 | Humans and Resources | 18 |
| 105 | KEY TERMS: Mix and Match | 18 |

### Investigating Ecosystems

| | | |
|---|---|---|
| 107 | Sampling Populations | 18 |
| 109 | Quadrat Sampling | 18 |
| 110 | Quadrat Based Estimates | 18 |
| 111 | Sampling a Leaf Litter Population | 18 |
| 113 | Transect Sampling | 19 |
| 117 | Mark and Recapture Sampling | 19 |
| 117 | Sampling Animal Populations | 19 |
| 118 | Indirect Sampling | 19 |
| 119 | Monitoring Water Quality | 20 |
| 121 | Radio Tracking | 20 |
| 122 | Classification Keys | 20 |
| 123 | Keying Out Plant Species | 20 |

# Contents

| Page | Title |
|---|---|
| 124 | KEY TERMS: Mix and Match ......... 20 |

## Land Water

| Page | Title |
|---|---|
| 127 | The Importance of Plants ............ 20 |
| 128 | Global Human Nutrition .............. 21 |
| 129 | The Green Revolution ................ 21 |
| 131 | Cereal Crop Production ............. 21 |
| 133 | Chemical Pest Control ............... 21 |
| 134 | Pesticide Resistance ................. 22 |
| 135 | Integrated Pest Management ..... 22 |
| 137 | Soil Degradation ....................... 22 |
| 139 | Reducing Soil Erosion ............... 22 |
| 140 | The Impact of Farming .............. 23 |
| 141 | Agricultural Practices ................ 23 |
| 143 | Forestry .................................... 23 |
| 145 | Managing Rangelands ............... 23 |
| 147 | Reserve Lands .......................... 23 |
| 149 | City Planning ............................ 24 |
| 151 | Transportation .......................... 24 |
| 152 | Environmental Remediation ....... 24 |
| 153 | Mining and Minerals .................. 24 |
| 154 | Globalization ............................. 24 |
| 155 | Ecological Impacts of Fishing .... 25 |
| 157 | Fisheries Management .............. 25 |
| 159 | KEY TERMS: Mix and Match ..... 25 |

## Energy

| Page | Title |
|---|---|
| 161 | Using Energy Transformations ... 25 |
| 162 | Non-Renewable Resources ....... 26 |
| 163 | Coal .......................................... 26 |
| 165 | Oil ............................................. 26 |
| 167 | Oil Extraction ............................ 26 |
| 170 | Environmental Issues of Oil Extraction ........... 27 |
| 171 | Nuclear Power .......................... 27 |
| 173 | Renewable Energy .................... 27 |
| 174 | Wind Power .............................. 28 |
| 175 | Hydroelectricity ......................... 28 |
| 177 | Solar Power .............................. 28 |
| 179 | Geothermal Power .................... 28 |
| 180 | Ocean Power ............................ 28 |
| 181 | Biofuels .................................... 28 |
| 182 | Current and Future Energy Demands ........ 29 |
| 183 | Energy Conservation ................ 29 |
| 185 | KEY TERMS: Mix and Match ..... 29 |

## Pollution

| Page | Title |
|---|---|
| 188 | Types of Pollution ..................... 29 |
| 189 | Water Pollution ......................... 29 |
| 191 | Sewage Treatment .................... 30 |
| 192 | Waste Management .................. 30 |
| 193 | Reducing Waste ........................ 30 |
| 194 | Plastics in the Environment ...... 30 |
| 195 | Atmospheric Pollution ............... 30 |
| 197 | Cities and Climate .................... 31 |
| 198 | Acid Rain .................................. 31 |
| 199 | Noise Pollution ......................... 31 |
| 200 | Toxicants in the Home .............. 31 |
| 201 | Health Effects of Pollution ......... 31 |
| 202 | The Economic Impact of Pollution ........... 32 |
| 203 | *Exxon Valdez* Oil Spill .............. 32 |
| 204 | Niger Delta Oil .......................... 32 |
| 205 | Deepwater Drilling .................... 32 |
| 206 | *Deepwater Horizon* Oil Spill ..... 32 |
| 209 | Oil Spills and Wildlife ................ 33 |
| 211 | Cleaning Up Oil Spills ............... 33 |
| 213 | Effects of Nuclear Accidents ..... 33 |
| 216 | Bhopal Disaster ........................ 33 |
| 217 | The Role of Legislation ............. 34 |
| 218 | KEY TERMS: Mix and Match ..... 34 |

## Global Change

| Page | Title |
|---|---|
| 220 | Models of Climate Change ........ 34 |
| 221 | Global Warming ........................ 34 |
| 223 | Biodiversity and Global Warming ........ 34 |
| 225 | Ice Sheet Melting ...................... 34 |
| 226 | Global Warming and Agriculture ........ 34 |
| 227 | Temperature and Distribution of Species ........ 35 |
| 228 | Ocean Acidification ................... 35 |
| 229 | Carbon Trading ......................... 35 |
| 230 | Carbon Capture and Storage .... 35 |
| 231 | Stratospheric Ozone Depletion . 35 |
| 233 | Loss of Biodiversity .................. 35 |
| 234 | Tropical Deforestation .............. 36 |
| 235 | The Impact of Alien Species ..... 37 |
| 236 | Endangered Species ................. 37 |
| 237 | Conservation of African Elephants ........ 37 |
| 238 | *In-situ* Conservation ................ 37 |
| 239 | *Ex-situ* Conservation ............... 37 |
| 241 | Conservation and Sustainability ........ 37 |
| 242 | Saving the Black Robin ............ 38 |
| 243 | The Sixth Extinction .................. 38 |
| 245 | KEY TERMS: Mix and Match ..... 38 |

## The Earth's History (page 11)

1. Stratigraphic record: Sequential layers of rocks in which environmental and biological changes are reflected in the changes in rock type and fossil assemblages.

2. (a) The record of the Earth's history is based on assigning a geological time period to a rock/fossil type (i.e. biostratigraphy).
   (b) Interpretation of strata can be complicated by unconformities in the strata (gaps in depositional history), disturbance of strata through uplift, tilting, and folding, and poor preservation of fossils.

3. Radiometric dating provides the means to reliably obtain the absolute (rather than the relative) age of rocks and other geological features, including the age of the Earth itself, and can be used to date a wide range of natural and man-made materials.

4. (a) Any of: horse, elephant, pig, numerous dinosaur groups, trilobites.
   (b) Any of: tuatara, coelacanth, gingkoes (ancient conifers).

5. Changes in the fossil record can help us understand how the environment has changed e.g. from marine to swamp to land, from tropical forest to cold plains. The fossil record of organisms provides information on the kinds of organism living in a particular environment. This information can be used to help confirm the type of environment that the organisms were living in e.g. woolly mammoths were likely to be living in very cold environments.

## Fossil Formation (page 13)

1. Decay

2. Marine organisms with hard parts tend to be relatively over-represented in the fossil record because shells (particularly thicker ones) are very durable and resistant to decay. Marine environments also provide frequent suitable environmental conditions for fossilization (rapid burial and protection from oxidation). In contrast, conditions for burial on land are often less favorable and organisms, if they do not have durable hard parts, decay quickly in most environments.

3. Fossils also provide clues to past environmental conditions, e.g. through changes in depositional environment. This information is often provided by isotope analysis.

## The Earth and the Sun (page 14)

1. B
2. C
3. B
4. D
5. (a) Solar heating is greatest in the tropics because of the Earth's curvature. Towards the poles, the Sun's rays impinge at an oblique angle and must pass through a greater thickness of atmosphere to reach the ground.
   (b) The Earth's atmospheric and ocean circulation (and therefore climate) arise as a consequence of this uneven heating. High levels of solar radiation at the tropics heat the lower atmosphere there causing air to rise and move to higher latitudes where it cools and sinks.

## The Earth's Crust (page 15)

1. The continental crust is much older and (related to this) it is not recycled within the Earth to the extent that the oceanic crust is.
The continental crust comprises many types of rocks (igneous, metamorphic, and sedimentary) but is largely granitic, whereas the oceanic crust is composed mainly of dense basalt-rich lavas and related rocks.

2. Because it is more dense, the oceanic crust is less elevated than the lighter continental crust and is of a relatively uniform thickness because it is constantly being recycled in the Earth through the spreading ridges. The continental crust is lighter, and therefore "floats" higher in the mantle and is variable in thickness as it is subjected to uplift and erosion.
**Further explanation**: The more mass there is above sea-level, the deeper the crust must extend down in support (a concept known as isostasy). Hence the continental crust is thickest beneath young mountain belts (where uplift is occurring). In contrast, the thin, dense oceanic crust reaches its equilibrium point well below dry land.

3. The Earth's crust is described as a dynamic structure because it is constantly undergoing change through ocean formation (new crust), mountain building, volcanism, and earthquakes, as well as compositional changes associated with rock cycle.

## Plate Boundaries (page 16)

1. (a) Convergent boundary: plates move towards one another; one plate slides under the other and crust is destroyed in the zone of subduction. Examples: the Andes, the Himalayas, the Pacific Ring of Fire.
   (b) Divergent boundary: plates move apart and new crust forms as the mantle rises and partially melts to form basaltic magma. Example: the Mid-Atlantic Ridge.
   (c) Transform boundary: plates slide past each other. Since crust is neither created nor destroyed they are sometimes called conservative plate boundaries. Example: San Andreas Fault, California (Pacific Plate and North American Plate), Alpine Fault, NZ (Pacific Plate and Indo-Australian Plate).

2. (a) Convection: Cooling and then heating of mantle material causes it to descend and then rise (hotter material is less dense). This provides the energy and movement for crustal movements.
   (b) Gravity pulls cooler, denser oceanic crust down into the mantle at subduction zones. As with convection, gravity provides some of the energy for plate movement.
   (c) Mantle plumes are narrow columns of very hot mantle material which rise to form hotspots associated with sustained island arc volcanism. **Extra note**: Although mantle plumes rise more-or-less independently of plate movements, they provide a continuous supply of magma in a fixed location, and may play a role in continental rifting and the formation of ocean basins.

3. (a) Mountain building: Convergent
   (b) Subduction: Convergent
   (c) Creation of new ocean floor: Divergent
   (d) Island arc: Convergent

4. Faults are cracks in the Earth's crust across which there has been movement. A sudden slippage or movement along a fault causes an earthquake; a sudden release of energy in the Earth's crust that creates seismic waves.

5. The Earth's plate boundaries are dynamic zones, where plates may converge, diverge, or move past each other. Subduction zones are associated with mountain building and volcanism while the stresses associated with plate movements are released as earthquakes.

6. During soil liquefaction, soil temporarily loses its strength, transforming from a solid to a liquid. In this state, it is no longer able to support the weight of structures built on it causing them to collapse.

## Lithosphere and Asthenosphere (page 18)

1. (a) The lithosphere comprises the crust and the upper part of the mantle. It is rigid and solid. The lithosphere is divided into 7 major plates and 12 smaller ones.
   (b) The asthenosphere is a semi-plastic layer that lies below the lithosphere. It is relatively thin and deforms slowly.

2. The lithosphere-asthenosphere boundary (LAB) can be detected as a thermal boundary (point where temperature changes rapidly). After this boundary the temperature

remains relatively steady even with increasing depth. Measuring the speed of seismic waves also provides evidence of the LAB. Seismic waves tend to travel more slowly through the asthenosphere than through the lithosphere.

3. Oceanic lithosphere is more dense than continental lithosphere. At convergent plate boundaries it tends to subduct into the mantle and be melted and recycled. Continental lithosphere, being less dense, tends to ride over the oceanic lithosphere and so remains relatively unchanged.

## Volcanoes and Volcanism (page 19)

1. (a) Andesitic
   (b) Rhyolitic
   (c) Basaltic

2. C

3. B

4. Shield volcanoes are formed from eruptions of fluid basaltic lava. This flows down the side of the volcano and spreads out, forming a volcano with a large basal radius and shallow slopes. Strato-volcanoes tend to be formed from andesitic lava that is less fluid than basaltic lava. The lava builds up around the vent of the volcano before flowing down the lower slopes. Strato-volcanoes have a shallow gradient on the lower slopes and a steeper gradient on the upper slopes.

5. The Hawaiian islands formed from a hotspot, a plume of basaltic magma that is apparently not related to any plate boundaries. As the Pacific plate moved over this hotspot in a north-westerly direction, new volcanoes were formed.

6. The Pacific Ring of Fire is located around the subducting edges of the Pacific and Nazca plates. Along the plates overriding these subduction zones are many volcanoes and thermal areas. In fact there are more volcanoes along the Ring of Fire than anywhere else on Earth.

7. Volcanic eruptions release large quantities of fine particles (ash) and gases into the atmosphere. The gases can contribute to the global warming effect, while the ash can block sunlight reaching the surface and so lower global temperature.

## The Rock Cycle (page 21)

1. Igneous rocks are volcanic in origin (they solidify from volcanic magma) although their composition varies widely. Examples include granite, obsidian, basalt, and rhyolite. Sedimentary rocks form where sediments accumulate and become compressed into rock. The sediments may originate from rocks of any type. Examples include mudstone, shale, loess, sandstone, breccia, greywacke, conglomerate, and limestone.
Metamorphic rocks result when pre-existing rocks (e.g. sedimentary rocks) are transformed by heat and pressure into a rock with a new texture and composition. Examples include marble, slate, schist, quartzite, and gneiss.

2. **Weathering** refers to the physical, biological, and chemical alteration of surface rocks (*in situ*). Over time, weathering breaks up and shapes rocks. **Erosion** is a collective term for the processes by which soil and rock material are loosened, dissolved, and (importantly) moved from any part of the Earth's surface. Weathering makes rock material more susceptible to erosion and erosion produces sediments, which may potentially give rise, through processes in the rock cycle, to other rock types.

3. The processes of the rock cycle are too slow to replenish minerals at the rate at which they are being extracted.

## Soil Textures (page 22)

1. A loam is a soil that contain a mixture of 40% sand, 40% silt, and 20% clay. Changes to these percentages are reflected in the names given to the loams. For example a sandy loam contains a greater percentage of sand than a silt loam.

2. – Soil sample 1:
   %Sand: 60    %Silt: 18    %Clay:22
   Soil type: Sandy clay loam
   – Soil Sample 2:
   %Sand: 72    %Silt: 15    %Clay:13
   Soil type: Sandy loam

3. Loamy soils provide a greater consistency of texture over a larger range of conditions than other loams. This provides more consistent air and water supply to a plant's roots and so provides a more favorable environment for plant growth. Loamy soils also tend to accumulate organic matter (humus) more effectively than other soil types.

## Soil and Soil Dynamics (page 23)

1. Weathering breaks down the parent rock to form a regolith, which overlies the bedrock. The properties of a soil arise, in part, from the result of this initial weathering process.

2. Climate, rock type, and topography separately and in combination influence soil properties. While rock type will determine the fundamental properties of a soil (e.g. its mineral content), climate (e.g. temperature, rainfall) regulates soil development by influencing chemical and physical weathering rates, as well as the influencing vegetation type and the soil biota. Topography influences soil development by affecting how well a soil retains moisture and is subject to erosion.

3. Soil organisms play an important role in soil development through their activities and (as a result of death and decay) by adding organic matter. A soil does not develop a mature horizon structure until it has an active biota contributing to it. Soil organisms incorporate organic matter into the mineral component of the soil, providing aeration and allowing moisture to penetrate. This improves soil texture and leads to deeper, humus-rich A horizon.

4. D

5. A

6. (a) Fertility: Organic content
   (b) Water-holding capacity (WHC): Soil texture (more clay and silt sized particles = higher WHC) and proportion of organic matter (more organic matter = greater WHC)

7. (a) Arctic soils: low temperatures slow the decomposition of organic matter so that accumulates on the soil surface.
   (b) Desert soils: The arid environment and (consequent) lack of vegetation leads to a paucity of organic matter and an impoverished soil biota. This limits the development of the A horizon and, as a consequence, poor vertical development.

## The Atmosphere and Climate (page 25)

1. Important roles of the atmosphere (two of):
   – Protects life by absorbing UV radiation.
   – Protects life by reducing temperature extremes between day and night.
   – Keeps the Earth warm (greenhouse effect).
   – Provides a breathable mix of gases for sustaining life.
   – Distributes heat around the globe.
   – Is the source of the Earth's climate.

2. Atmospheric circulation is driven by the (uneven) solar input (the Sun's radiant energy), which results in air rising at lower latitudes and sinking at the poles.

3. (a) Troposphere: The zone in which all weather occurs. Mixed horizontally and vertically.
   (b) Stratosphere: Temperature stable at -60°C to 20 km then increases due to UV absorption.

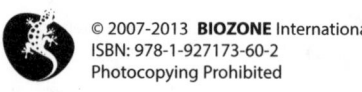

© 2007-2013 **BIOZONE** International
ISBN: 978-1-927173-60-2
Photocopying Prohibited

4. (a) The prevailing wind is the wind direction that is most frequent across a particular region.
   (b) The prevailing wind of a region is reflected in some of the physical and biological characteristics of the environment. For example, in polar and desert regions, dunes and drifts develop characteristic shapes as material is continually pushed in one direction. Prevailing winds also sculpt vegetation so that growth is restricted to the leeward side.

5. The ITCZ is also called the doldrums because it is a equatorial belt of calm between the two belts of trade winds. In this area, there are frequent periods when the winds disappear, becalming sailing vessels for days or weeks.

## Variation and Oscillation (page 27)

1. Winds (since they are just moving air molecules) are subjected to the Coriolis effect, which deflects wind flows to the right in the Northern Hemisphere and to the left in the Southern Hemisphere. This creates the trade winds and prevailing westerlies in each hemisphere.

2. (a) Movement a cold front:

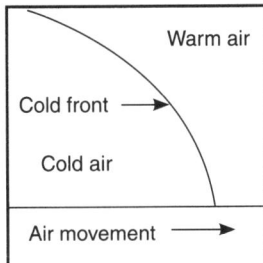

   (b) Air movement in a Southern Hemisphere cyclone and Northern Hemisphere hurricane:

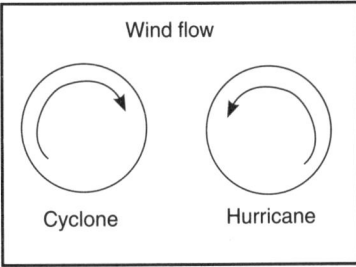

3. El niño conditions occur when air pressure over the Indian Ocean, Indonesia, and Australia rises. Warmer waters extend eastward and block oceanic upwellings along the coast of the Americas. Warmer air rises over South America and descends over Australia (the reverse of normal conditions).

4. (a) In an El Niño year, the western coast of South America (Peru and Chile) receives more rain that usual. The normal climatic conditions are reversed, with warm water spreading from the west Pacific and the Indian Ocean to the east Pacific. It takes the rain with it, as warmed air rises near Peru, causing higher rainfall in the deserts there.
   (b) In an El Niño year, there is a rise in air pressure over the Indian Ocean, Indonesia, and Australia which brings warmer, drier weather to these areas. Australia and Indonesia suffer droughts and frequent fires as a result.

## Ocean Circulation and Currents (page 29)

1. The thermohaline circulation (thermo=heat, saline=salt, together determine the density of seawater) is driven by the cooling and sinking of water masses to great depth in the North Atlantic. Atlantic waters are saltier and more dense that those of the Pacific. In the north, they cool and sink, flowing south to penetrate the Pacific and Indian oceans, before returning as warm surface ocean currents in the South Atlantic. The polar oceans are the sources of the cold dense bodies of water that drive thermohaline circulation.

2. The thermohaline circulation transports energy (in the form of heat) around the globe. As such, the state of the circulation has a large impact on the Earth's climate. It has an important role in supplying heat to the polar regions and thus in regulating the amount of sea ice in these regions. Disruption of the thermohaline circulation is one possible consequence of global warming.
   **Further explanation:** Some climate models predict that large influxes of low density meltwater from the Greenland ice sheet could lead to a disruption of deep water formation and subsidence in the extreme North Atlantic. This could trigger localized cooling in regions currently warmed by the North Atlantic Drift (part of the thermohaline circulation (THC)), e.g. Europe and England. A slowdown of the THC could have other climate consequences too, such as an increase in major floods and storms, a collapse of plankton stocks, warming or rainfall changes in the tropics or Alaska and Antarctica (including those from intensified El Niño effect), and more frequent and intense El Niño events.

3. A significant reduction in the Earth's albedo would result in the Earth reflecting less of the Sun's radiant energy (absorbing more energy), which would cause further warming and (in turn) more melting. This ice-albedo positive feedback loop would accelerate the rate of changes in global temperatures (warming, but also potentially localized cooling; see the answer to question 2 above).

4. The deep water circulation is driven by differences in the temperature and density of large water masses moving at great depth. Surface circulation is driven by winds and modified by the Coriolis effect.

5. (a) Antarctic circumpolar current:   E
   (b) Peru current:   D
   (c) South Pacific gyre:   C
   (d) North Atlantic gyre:   A
   (e) South Atlantic gyre:   B

6. Both atmospheric and surface circulation patterns transport heat from equatorial regions to the poles.

7. (a) Upwelling returns nutrients to surface waters which supports the growth of plankton (thus supporting food webs).
   (b) In an El Niño year, a stable high pressure system develops over Australia and warm water extends deeper and flows eastwards towards the west coast of South America. This warm water blocks the upwelling of cold nutrient-rich water that normally supports fish stocks along the west coast of South America.

## Global Water Resources (page 31)

1. Surface water is free water that exists at the surface in lakes, streams, and rivers. Groundwater is water located beneath the ground surface in soil pore spaces and in the fractures of rock formations (as aquifers).

2. (a) Some extensive aquifers are never-the-less non-renewable because their recharge times are so long (thousands of years).
   (b) Long term viability of an aquifer is affected by recharge rates (rate at which the aquifer is replenished), and rates of extraction (e.g. for irrigation). Low replenishment rates and high rates of withdrawal would make an aquifer non-viable in the long term.

3. (a)-(c) Any of the following in any order:
   – Withdrawal of groundwater
   – Desalination plants
   – Transport water in from other areas
   – Build dams and reservoirs to store runoff
   – Improve the efficiency of water use (including using gray water for irrigation)
   – Treatment of large volumes of waste water

4. Water stored in ice has an important role in regulating climate; by increasing the Earth's albedo (reflectivity), they have a cooling effect.

5. (a) Delta regions are very fertile because they are associated with the floodplain region of rivers and receive the regular deposits of silts and sediments when the river floods. These regular deposits restore the soils, renew nutrient levels, and build new land.
   (b) Flood control schemes interfere with the normal seasonal flooding regime of the river, preventing the annual resuppply of nutrients and sediments from upstream areas. This leads to depletion of the floodplain area and problems with salinization as the soils miss out on their annual flushing.

6. The Earth's major water resources (rivers, lakes, and aquifers), being reliable sources of potable water are associated with major concentration of the world's populations. Examples include the Nile floodplain, areas supplied by the Ogallala aquifer, the Mississippi, and the Murray-Darling basin. For the same reasons, they are also areas associated with high biodiversity. This creates conflict as the native plants and animals associated with the region must compete for limited resources with an expanding human population.

---

## Water and People (page 33)

1. (a) Water is less dense as a solid than a liquid (so it floats), it is polar (so it is able to dissolve many substances), forms strong intermolecular bonds, and has a very high boiling point. It is colorless, tasteless and odorless.
   (b) Water is important for life because of its ability to dissolve many substances, form strong intermolecular bonds, and remain liquid at high temperatures and low pressures. These properties make it the ideal solvent for carrying out the chemical reactions within a cell.

2. Water supply can be limited in some countries by a lack of rain fall or stable waterways such as large rivers or lakes. An imbalance in the distribution of rains or waterways can limit supplies of water in some parts of a country. Unstable governments and/or poor water distribution systems are major causes for the lack of water in many countries.

3. (a) Pipelines should not leak. Irrigation needs to be targeted to specific areas (no over spray). Drip irrigation and underground irrigation reduces evaporation of water and delivers it direct to the plant's roots.
   (b) Pipelines should not leak. Evaporation of stored water needs to be reduced. Water should be running only when be used (e.g. taps should be turned off when brushing teeth). Used water should be piped into gray-water tanks for use outdoors (non-drinking water).
   (c) Pipelines should not leak. Water used for washing can be used in gray-water systems for outdoor use (e.g. watering the garden).
   (d) Minimum amount of water for cooking should be used. Used cooking water can be reused for outdoor use (e.g. watering the garden).

4. Water is a source of conflict because it is so essential to life that its acquisition is a primary human need. Not only is water needed for drinking but for many other purposes including cooking and sanitation. Water supplies running from one country into the next may be a source of conflict if the first country is drawing too much water or polluting the water course so that the second country can not use it. Conflict can also be within states of the same country such as between Colorado, New Mexico and Texas, all of whom claim a share of the water flowing through the Rio Grande.

5. C
6. B

---

## Water and Industry (page 35)

1. The USA is an industrialized country, where much of the population is urbanized and agriculture is mechanized. This is reflected in its water use, which is directed primarily towards power production and agriculture. In contrast, China has a huge rural population (most of its rice production for example is for internal consumption) and irrigation (for agriculture) accounts for the largest use of water.

2. Water availability is a critical resource issue this century for a number of connected reasons. Primarily, pressure on water (especially potable water) resources is increasing as the world's population expands, especially as it expands in restricted areas such as large cities. This population pressure creates further problems with water quality associated with pollution of existing water supplies. Adding to the pressure on existing water supplies is the depletion of previously reliable supplies as a result of climate changes (persistent droughts) or the historical effects of heavy withdrawals (Ogallala aquifer). Climate change has other effects too, quite apart from regional droughts. For example, extensive flooding in some areas results in pollution of potable water and creates shortages of useable water. Extreme weather events are predicted to become more frequent under a regime of climate warming, which makes the water resource issue even more pressing.

3. (a) Crop irrigation: The efficiency of crop irrigation can be substantially improved with the use of improved protocols and equipment, despite what can be high initial costs. Any of the following could be discussed:
   – Lining canals bringing water to irrigation ditches.
   – Using soil moisture detectors to determine when crops need water and then irrigating crops only when necessary.
   – Polyculture.
   – Switching to water-efficient crops.
   – Irrigating with treated urban waste water.
   – Importing water-intensive crops and meat.
   – Improving irrigation systems, e.g. using center-pivot low pressure sprinkler systems and drip irrigation, or using time controlled valves on standard gravity flow systems.
   (b) Industrial water use: There are many possible and viable options for reducing industrial water use including:
   – Redesigning manufacturing processes to use recycled water.
   – Manufacturing from scrap rather than virgin ore (e.g. aluminium).
   – Install water efficient devices and appliances (dual flush toilet, tap timers).
   – On production lines fit movement detectors on water sprays so water is used only when products pass them.
   – Use trigger gun hoses. A running hose wastes 1000 litres per hour.
   – Educate employees on water efficiency.
   – Fix leaks quickly.
   – Install a rainwater tank to capture rainwater run off. This can be used for many purposes around the workplace such as a commercial garden or in work processes.
   – For pipelines that carry manufactured product and need periodic cleaning, consider pipe pigs, air or inert gas pulses to reduce the amount of water used for washing.
   – Bleed off cooling towers by conductivity (regulating salts in cooling water).
   – Recycle wastewater. Often wastewater can be used directly or with little treatment on a commercial garden or in the manufacturing process.
   – Use a broom instead of a hose to sweep down production areas and pathways (dry areas).
   – Recycle cooling water through a cooling tower rather than a single pass through to the sewer system.
   – Insulate hot water pipes. This avoids wasting water and power while waiting for hot water to flow through pipes.

4. Students can refer to the cited example of the Aswan High Dam or any other local or international example of their choice. Environmental problems associated with damming a large river include problems associated with the dam itself and those associated with dam operation. These include:
Environmental impacts of the dam itself
   – Imposition of a reservoir in place of a river valley (resulting in habitat loss).
   – Changes in downstream morphology of riverbed, delta, or coastline due to altered sediment loads (increased erosion).
   – Changes in downstream water quality: effects on river temperature, nutrient load, turbidity, dissolved gases, concentration of heavy metals and minerals.
   – Reduction of biodiversity as a result of blocking the movement of organisms (e.g. salmon) and because of the changes described above.
Environmental impacts of dam operation:

- Changes in downstream hydrology: alteration of seasonal and total flow volumes, changes to extreme high and low flows, short-term fluctuations in flows (sometimes hourly).
- Changes in downstream morphology caused by altered flow patterns.
- Changes in downstream water quality caused by altered flow patterns.
- Reduction in riverine/riparian/floodplain habitat diversity, especially because of elimination of floods.

Economic problems associated with damming a large river are associated with them not fulfilling the economic promises made for them. These can include:
- Not producing the power output promised.
- Silting up earlier than expected.
- Increasing expense, e.g. for raw materials and to meet mitigation requirements.
- Delays due to public opposition and the time taken to build the dam.
- Relocation of communities and the social and economic costs associated with this.
- Increased disease risk associated with dams providing habitat for parasites.
- Inflexibility of hydrodams as power source (need to be able to predict electricity demands far in to the future).
- Dams are an important contributor to national debt.
- Economic inequalities, which can result in the promised benefits (e.g. irrigation) preferentially benefitting the already advantaged.

5. Student's own research. Answers will vary depending on the region in which they live.

## KEY TERMS: Mix and Match (page 37)

Atmosphere (AA), Convergent boundary (CC), Core (G), Coriolis effect (Z), Crust (F), Divergent boundary (K), ENSO (BB), Fault (J), Groundwater (I), Igneous (S), Intertropical convergence zone (W), Lithosphere (N), Loam (C), Mantle (P), Metamorphic (Q), Plate boundary (R), Plate tectonics (O), Radiometric dating (L), Rock cycle (M), Sedimentary (A), Shield volcano (E), Soil (D), Soil horizon (V), Stratosphere (T), Strato-volcano (H), Thermohaline circulation (U), Transform boundary (X), Tricellular model (B), Troposphere (Y).

## Components of an Ecosystem (page 40)

1. (a) B Population
   (b) C Ecosystem
   (c) D Physical factor
   (d) C Ecosystem
   (e) A Community

2. The **biotic factors** are the influences that result from the activities of living organisms in the community whereas the **abiotic** (physical) **factors** are the influences of the non-living part of the community, e.g. climate.

## Factors Affecting Biome Distribution (page 41)

1. The Sun's rays hit the tropics almost perpendicular to the Earth's surface so that their energy is concentrated in a small area. This causes the tropics to be very hot. The heat at the surface causes air to rise. This air carries with it a large amount of water vapor. This condenses at altitude and forms rain. This process is particularly intense in the tropics because of its location at the convergence of two large air cells. (Hadley cells, one to the south and one to the north).

2. The Northern Hemisphere contains a lot more land than the Southern Hemisphere. This causes changes in the distribution of biomes, with the Southern Hemisphere being more influenced by the sea. It also gives more of land in the north for the development of a wider range of biomes.

## World Distribution of Biomes (page 42)

1. These are rainshadow areas: dry areas in the leeward side of mountains in the path of rain-bearing winds. Much of the precipitation is dropped at high altitude in the mountain ranges as snow and ice, so there is little precipitation in lowland areas adjacent to mountains

2. Most of the natural extent of temperate forest is mid-latitude with a reasonably equable climate and moderately high, evenly distributed rainfall. These factors make the forest area ideal for settlement and agriculture. Consequently, much of the original forest has now been cleared.

3. Northern extent of boreal forest limited by low temperatures and short daylight hours for half of the year (= short growing season), and the presence of a permanently frozen ground layer which prevents deep rooting (required by larger trees).

## Effect of Temperature on Biomes (page 44)

1. The distribution of biomes about the globe matches the conditions of rainfall and temperature. Similar biomes are therefore found in similar conditions. e.g. tropical desert biomes are found in areas of high temperature and low rainfall while tropical rainforests are found in areas of high temperatures and high rainfall.

2. Biomes are not evenly distributed around the globe because rainfall and temperature are not even about the globe. Mountains cause rain on one side and rain shadows on the other and affect air temperature. Large bodies of water such as oceans and large lakes can also affect temperature and rainfall. As these are not evenly distributed about the globe neither are biomes.

3. Landscape influences climate by affecting the air travelling over it by either adding or removing water and raising or lowering the temperature. High mountains deflect wind and remove water from the air as rain or snow, creating dry air and elevated temperatures on their leeward slopes. Bodies of water tend to remain at the same temperature and so modify nearby land temperatures, keeping them more even through both night and day and through the seasons.

4. The energy received from the Sun depends upon angle of the surface of the Earth in respect to the Sun. The angle the surface of the Earth presents to the Sun changes from almost ninety degrees at low latitudes (the equator) to almost zero at higher latitudes (the poles). This causes the energy from the Sun to be spread out over a greater surface area, so that a square kilometer of land at the poles receives much less solar energy than a square kilometer of land at the tropics.

## Physical Factors and Gradients (page 45)

1. Climate refers to longer term, usually broad scale weather patterns in a region. Microclimate refers to climatic variation in a very small area or in a particular habitat. This can vary depending on shelter and aspect, as well as the influence of objects in the environment. It often refers to the immediate climate in which an organism lives.

2. High humidity underground, in cracks, under rocks.

3. In a crack or crevice, in a burrow underground, in spaces under rocks.

4. An animal unable to find suitable shelter would undergo heat stress, dehydration, and eventually die.

5. High humidity enables land animals to reduce their water loss due to evaporation. This in turn reduces their demand for (and dependence on) drinkable water.

6. At night, temperature drops and humidity increases (to the point where condensation may occur; this is a source of valuable water for some invertebrates).

## Physical Factors in a Forest (page 46)

1. Environmental gradients from canopy to leaf litter:
   (a) Light intensity: decreases.
   (b) Wind speed: decreases.
   (c) Humidity: increases.

2. Reasons why factors change:
   (a) Light intensity: Foliage above will shade plants below, with a cumulative effect. The forest floor receives light that has been reflected off leaf surfaces several times, or passed through leaves.
   (b) Wind speed: Canopy trees act as a wind-break, reducing wind velocity. Subcanopy trees will reduce the velocity even further, until near the ground the wind may be almost non-existent. An opening in the forest canopy (a clearing) can expose the interior of the forest to higher wind velocities.
   (c) Humidity: The sources of humidity (water vapor) are the soil moisture, leaf litter, and the transpiration from plants. Near the canopy, the wind will carry away moisture-laden air. Near the forest floor, there is little wind, and humidity levels are high.

3. The color of the light will change nearer the forest floor. White light (all wavelengths) falling on the canopy will be absorbed by the leaves. Reflected light in the green wavelength bounces off the leaves and passes downward to lower foliage and the forest floor.

## Stratification in a Forest (page 47)

1. Stratified forests have distinctive layers each with its own microclimate. As a result, the diversity of habitats is much greater than found in a forest with less vertical structure. For example, trees in the canopy layer receive full sun, but must be adapted to cope with higher winds. Plants in the ground layer must be shade adapted as very little sunlight filters down to the ground, but they do not need to be adapted to cope with wind.

2. Removal of emergents and large canopy trees opens up gaps in the canopy, altering the physical factors within the forest. More sunlight reaches lower levels of the forest, trees previously under the canopy trees may receive more precipitation, and trees in the subcanopy may be exposed to more wind. Removal of canopy trees may give other species an opportunity to flourish, as a result the composition of the forest may change.

## Physical Factors on a Rocky Shore (page 48)

1. Environmental gradients:
   (a) Salinity: Increases from LWM to HWM
   (b) Temperature: Increases from LWM to HWM
   (c) Dissolved oxygen: Decreases from LWM to HWM
   (d) Exposure: Increases from LWM to HWM

2. Rock pools may have very high salinity due to evaporation after long exposure times without rain.

3. (a) Mechanical force of wave action: Point B will receive the full force of waves moving inshore, Point A will receive only milder backwash, Point C will experience some surge but no direct wave impacts.
   Surface temperature: Points A and B will experience greater variations in rock temperature depending on whether the tide is in or out, day or night, water temperature, wind chill. Point C is more protected from some of these factors and will not experience the warming effect of direct sunlight.
   (b) Microclimate.

## Physical Factors in a Small Lake (page 49)

1. Environmental gradients from water surface to bottom:
   (a) Water temperature: Decreases gradually until below the zone of mixing when there is a sharp drop.
   (b) Dissolved $O_2$: Oxygen at a uniform concentration until below the zone of mixing when there is a sharp drop, with very little oxygen at the bottom.
   (c) Light penetration: Decreases at an exponential rate (most light is absorbed near the surface).

2. (a) Prevents mixing of the oxygen-rich surface water with the deeper oxygen-deficient water (represents a thermal barrier).
   (b) Organisms (particularly bacteria) living below the thermocline use up much of the available oxygen. Decomposition also uses up oxygen.

3. (a) Heavy rainfall or inflow of floodwater from nearby river channels may cause a decline in salinity.
   (b) Evaporation from the lake concentrates salts and the conductivity will increase.

4. Physical gradients will govern what species will be found and where in a particular area, as determined by their specific tolerances to abiotic factors.

## Habitat (page 50)

1. An organism will occupy habitat according to its range of tolerance for a particular suite of conditions (temperature, vegetation and cover, pH, conductivity). Organisms will tend to occupy those regions where all or most of their requirements are met and will avoid those regions where they are not. Sometimes, a single factor, e.g. pH for an aquatic organisms, will limit occupation of an otherwise suitable habitat.

2. (a) Most of a species population is found in the optimum range because this is the zone where conditions for that species are best; most of the population will select that zone.
   (b) The greatest constraint on an organism's growth within its optimum range would be competition between it and members of the same species (or perhaps different species with similar niche requirements).

3. In a marginal niche, the following might apply:
   – Physicochemical conditions (e.g. temperature, current speed, pH, salinity) might be sub-optimal and create stress (therefore greater vulnerability to disease).
   – Food might be more scarce or of lower quality/nutritional value.
   – Mates might be harder to find.
   – The area might be more exposed to predators.
   – Resting, sleeping, or nesting places might be harder to find and/or less suitable in terms of shelter or safety.
   – Competition from other better-adapted species might be more intense.

## Ecological Niche (page 51)

1. (a) The realized niche could be regarded as flexible in that it represents the constrained limits of the potential or fundamental niche. In the absence (or reduction) of interspecific competition (or other population constraints such as predation), a species could expand the extent of its realized niche. This can occur when a species becomes an unwanted pest in a new environment. Similarly, if competition intensified, then the realized niche would contract.
   (b) Factors constraining the extent of the realized niche: competition, predation, parasitism and disease.

2. Interspecific competition, when intense, results in some overlap in resource use curves and selection will favor a contraction of the niche and species specialization (of resource use). This (favorable) result of this is a divergence in resource use curves. Intraspecific competition in contrast, acts to broaden niches because competing individuals are forced to exploit resources at the extremes of their tolerance range. Because the competing individuals are the same species (with the same resource requirements) specialization into different niches is generally not an option (within the constraints of the species biology). **Teacher's explanatory note**: Of course, there are a number of examples of ecologically and genotypically flexible species moving into new regions unoccupied by competitors. In these cases, character displacement within one species can result in speciation given the appropriate conditions of isolation and resource and niche availability. The finches of the Galàpagos are one example. In most ecological situations though, intense competition within a species when niches are limited forces a broadening of that species niche.

## Energy Inputs and Outputs (page 52)

1. Producers convert energy received from an inorganic source (usually sunlight) into a form that is accessible to consumer levels. Consumers depend on the energy stored in the chemical bonds of biological molecules (the fats, proteins, and carbohydrates of plant and animal tissues). They too transfer energy to other levels, but energy is lost with each transfer.

2. In a grazing food web, energy moves from producers (plants) to primary consumers (herbivores) and then to secondary consumers (carnivores). This chain of energy transfer can continue several times, but eventually ends. All these consumer groups provide energy to decomposer levels. In a detrital food web, producers provide energy as dead plant material, and the primary consumers are decomposer microorganisms such as bacteria and fungi. Energy flows back and forth between decomposers and detritivores but herbivores and carnivores do not feature.

3. Detritivores consume detritus (decomposing organic material), and in doing so they speed up decomposition by increasing the surface area available to decomposer bacteria. Decomposers (bacteria and fungi) also use detritus as an energy source but digestion is extracellular (enzymes are secreted in fungi or bound to the cell surface in bacterial cells). These enzymes break down the detritus into constituent molecules for absorption so the breakdown is more complete than is the case with detritivores.

## Photosynthesis (page 53)

1. Photosynthesis is the process in which inorganic carbon is fixed to form organic carbon molecules. It is also the process in which energy from the Sun is converted into chemical energy. These two process form the basis for life on Earth, providing the carbon and the energy required by all living things.

2. Leaves absorb blue and red light and reflect green light. The reflected light makes the leaf appear green.

3. The end product of photosynthesis is a six carbon sugar - glucose. This is used to form a number of other complex molecules - cellulose, starch and disaccharide sugars. These molecules are used for:
   - Cellulose: Major component of plant cell walls.
   - Starch: Storage of energy rich glucose molecules.
   - Disaccharides: Sugars, such as fructose, accumulate in fruit, attracting seed dispersers.
   - Glucose: The molecule used as fuel for cellular respiration.

4. (a) Oxygen is produced from the breakdown of water.
   (b) Carbon dioxide combines with hydrogen from water (via a number of chemical steps) to produce glucose.

## Measuring Primary Productivity (page 54)

1. LAI is an indirect measure of the capacity of the plant to intercept light, photosynthesize, and produce new plant biomass. High LAIs may be typical of high productivity, although this is not necessarily the case. Large trees may support a large biomass with relatively low rates of production of new biomass. Conversely, grasslands may have high rates of production despite cropping because they are not supporting a large, static biomass. The concept is more applicable in agricultural situations where there can be marked reductions in the (typical) LAI for a crop, e.g. through crop damage due to pests, and this reduces the primary productivity. Note: The LAI is the measure of the leaf area of a plant exposed to incoming light, expressed in relation to the ground surface area beneath the plant.

2. (a) The procedure outlined only gives an estimate of NPP because you cannot easily determine the biomass not collected due to consumption or death.
   (b) To express standing crop in kJ m-2 you would need to know the energy content of the plant material (per known mass unit measured, e.g. kg). Note: A calorimeter could be used to do this.
   (c) To calculate GPP you would need to know the energy lost in respiration, e.g. by measuring the $CO_2$ lost at night.

## Cellular Respiration (page 55)

1. (a) Glycolysis: cytoplasm
   (b) Krebs cycle: matrix of mitochondria
   (c) Electron transport chain: cristae (inner membrane surface) of mitochondria.

2. (a) ATP
   (b) ATP acts as a chemical energy transport molecule, transferring the energy gained from the breakdown of glucose to other biochemical pathways.

## Food Chains (page 56)

1. (a) The Sun.
   (b-d) Refer to the diagram below.

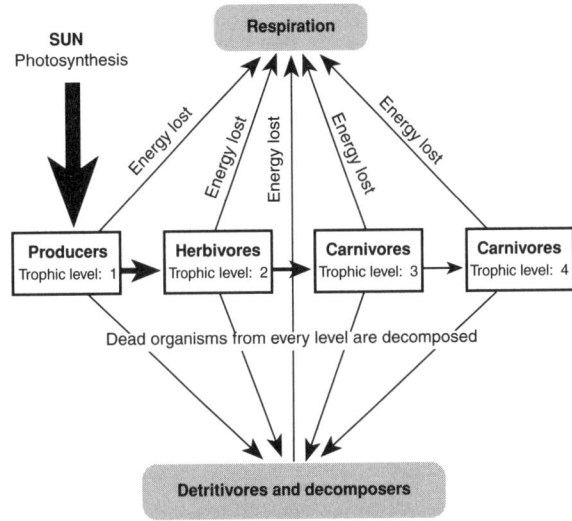

Some secondary consumers feed directly off decomposer organisms

2. (a) Each successive trophic level has less energy.
   (b) Energy is lost by respiration as it is passed from one trophic level to the next.

3. A food chain comprises a sequence of organisms, each of which is a source of food for the next. Food chains are organized according to trophic levels; the feeding levels that energy passes through as it proceeds through the food chain. Organisms are assigned a category according to the trophic level they occupy. Producers form the first trophic level, 1st order consumers (primary consumers) eat producers (i.e. herbivores), 2nd order consumers (secondary consumers) eat herbivores (i.e. they are carnivores) etc. Organisms may occupy more than one trophic level depending on their diet. Detritivores and decomposers obtain energy from all other trophic levels and are therefore not assigned a trophic level.

4. The kingfisher occupies trophic level 3 or 4 at different times (depending on the prey of choice).

## Food Webs (page 57)

1. Some food chain examples as below (there are others)
   (a) Algae → zooplankton → diving beetle
   (b) Algae → zooplankton → stickleback → pike
   (c) Macrophyte → great pond snail → herbivorous water beetle → stickleback → pike
   (d) Macrophyte → carp → pike
   (e) Algae → mosquito larva → Hydra → dragonfly larva → carp → pike
   (f) Macrophyte → herbivorous water beetle → carp → pike
   (g) Algae → zooplankton → *Asplanchna* → leech → dragonfly larva → carp → pike
   (h) Detritus → *Paramecium* → *Asplanchna* → leech → dragonfly larva → carp → pike

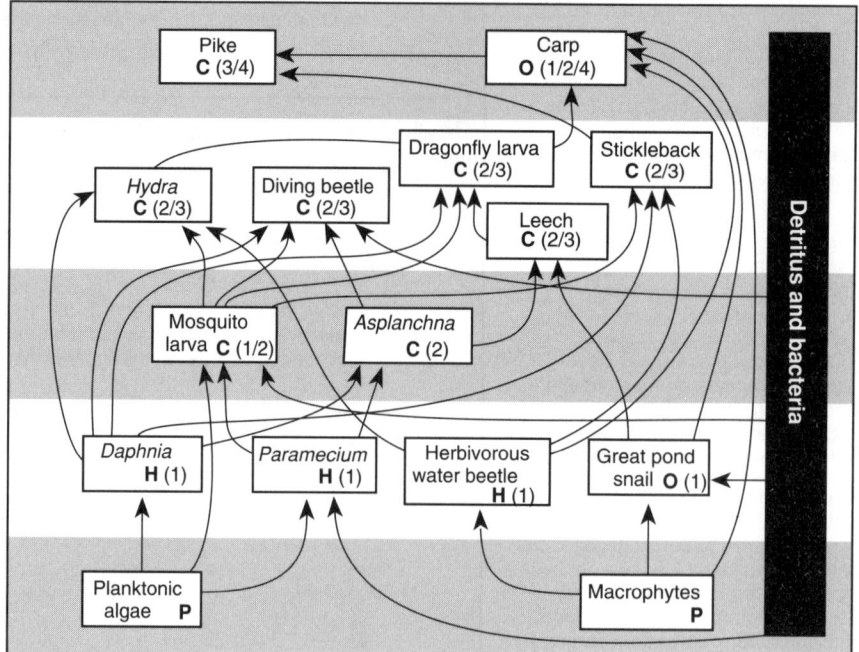

(i) Detritus → great pond snail → leech → dragonfly larva → carp → pike
(j) Detritus → *Paramecium* → mosquito larva → Hydra → dragonfly larva → carp → pike

2. (a-b) See diagram above.
Clarification: we apologize for the confusing notation used in this exercise. In the example given for mosquito larvae, the C2 refers to consumer level 2 (not carnivore). It occupies this position based on the assumption that it consumes *Paramecium* (as part of the plankton) which we have not included in its description.

## Energy Flow in an Ecosystem (page 59)

1. (a) 14,000
   (b) 180
   (c) 35
   (d) 100

2. Solar energy

3. A. Photosynthesis
   B. Eating/feeding/ingestion
   C. Respiration
   D. Export (lost from this ecosystem to another)
   E. Decomposers and detritivores feeding on other decomposers and detritivores
   F. Radiation of heat to the atmosphere
   G. Excretion/egestion/death

4. (a) 1,700,000 ÷ 7,000,000 x 100 = 24.28%
   (b) It is reflected. Plants appear green because those wavelengths are not absorbed. Reflected light falls on other objects as well as back into space.

5. (a) 87,400 ÷ 1,700,000 x 100 = 5.14%
   (b) 1,700,000 - 87,400 = 1,612,600 (94.86%)
   (c) Most of the energy absorbed by the producers is not used in photosynthesis. This excess energy which is not fixed is lost as heat (although the heat loss component before the producer level is not usually shown on energy flow diagrams). Note: Some of the light energy absorbed through accessory pigments widens the spectrum that can drive photosynthesis. However, much of accessory pigment activity is associated with photoprotection; they absorb and dissipate excess light energy that would otherwise damage chlorophyll.

6. (a) 78,835 kJ
   (b) 78,835 ÷ 1,700,000 x 100 = 4.64%

7. (a) Decomposers and detritivores
   (b) Transport by wind or water to another ecosystem.

8. (a) Energy remains locked up in the detrital material and is not released.
   (b) Geological reservoir:

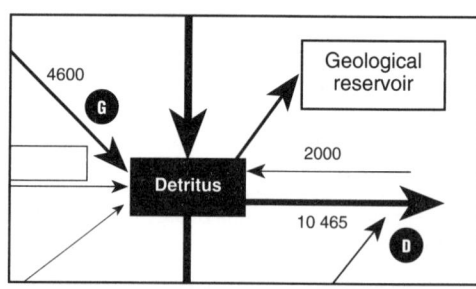

(c) Oil (petroleum) and natural gas, formed from the remains of marine plankton. Coal and peat are both of plant origin; peat is partly decomposed, and coal is fossilized.

9. (a) 87,400 → 14,000:
   14,000 ÷ 87,400 x 100 = 16%
   (b) 14,000 → 1600:
   1600 ÷ 14,000 x 100 = 11.4%
   (c) 1600 → 90:
   90 ÷ 1600 x 100 = 5.6%

## Production and Trophic Efficiency (page 61)

1. Net primary production refers to the net amount of producer biomass available per unit area to the next trophic level. Net primary productivity is the rate of producer biomass production or the producer biomass produced per area per unit time.

2. (a)-(c) and any of the following:
   – Amount and availability of light for photosynthesis. This is higher in the tropics.
   – Temperature. Higher temperatures are generally conducive to higher productivity.
   – Availability of water. Photosynthesis (and therefore productivity) will be limited when water is scarce.
   – Availability of nutrients. Nutrient limitations will limit plant growth and lower productivity.

3. (a)-(c) any of the following:
   – High diversity of grassland species contributing to ecosystem efficiencies (different tolerances and preferences so photosynthesis is maximized).
   – High root production typical of herbaceous species.
   – High producer turnover because continual cropping by herbivores keeps plants actively growing.
   – High rates of nutrient recycling so nutrient availability is not a factor in limiting productivity.
   – Savanna occurs in equatorial or near equatorial latitudes so high light and temperature help to increase rates of plant photosynthesis.
   – Continual supplies of nutrient from the dung of primary

consumers (grazing herbivores).

4. (a) The key factors limiting rates of primary production in terrestrial ecosystems are temperature and moisture; the productivity of tundra ecosystems is limited by low temperatures, while that of desert ecosystems is limited by moisture availability. Tropical rainforest ecosystems do not have these same limitations.
   (b) In aquatic systems, light and nutrient availability limit rates of production. The NPP of open ocean is low relative to coastal systems because of the low levels of nutrients. Nitrogen and phosphorus, in particular, are very low in the open ocean but higher in coastal systems which receive inputs from the land.
   Note: Although light may be limiting to productivity in the open ocean, tropical waters are less productive than one would predict from the higher light intensities there; low nutrient availability is the critical factor in this case.

5. (a) NPP of a particular crop can be maximized by reducing losses to pests, spraying to reduce losses to disease, maximizing light penetration and nutrient uptake by keeping competing plants to a minimum, ensuring adequate amounts of water and nutrients.
   (b) Livestock productivities can be maximized using the same principles: minimizing levels of disease and ensuring optimum nutrition, optimizing stocking densities to reduce stress, reducing energy losses by restricting excessive movement (smaller paddocks) and providing shelter belts or indoor housing in colder weather.

6. Deserts are often high light environments, but are limited by water and nutrient availability so productivity is low. Intensive horticultural land is provided with plentiful water and nutrients and crops are often kept in climate controlled environments to maximize temperatures and raise carbon dioxide levels. The result is very high productivity (maintained by continual high inputs of energy and matter).

## Ecological Pyramids (page 63)

1. (a) Number pyramid: Numbers of individual organisms at each trophic level.
   (b) Biomass pyramid: Weight (usually dry weight) of all organisms at each trophic level.
   (c) Energy pyramid: Energy content of all organisms at each trophic level.

2. Biomass or energy pyramids usually more accurately reflect the energy available to the next trophic level than pyramids of numbers. Pyramids of numbers can be misleading because a small number of producers may represent a large amount of biomass or energy.

3. Producers include the large trees. These have a large biomass and energy content per individual.

4. (a) 8690 → 142 = 8548 kJ = 1.6%
   (b) 142 → 12 = 130 kJ = 8.5%
   (c) Energy passed on from producers to primary consumers is less than the expected 10% because a lot of energy is diverted to the decomposers.
   (d) Decomposers
   (e) In a plankton community, turnover times (generation times of organisms) are very short and there is a lot of dead material both in the water and on the bottom. This provides a rich energy source to support a large biomass of decomposers.

5. The algae are reproducing at a high rate, but are being heavily cropped by the larger biomass of zooplankton.

## Species Interactions in Communities (page 65)

1. Savannah:

| Interaction | Zebra | Species B |
|---|---|---|
| Competition | – | Wildebeest (–) |
| Parasitism | – | Tick (–) |
| Predation | – | Lion (+) |

Redwood forest:

| Interaction | Redwood | Species B |
|---|---|---|
| Competition | – | Redwood (–) |
| Commensalism | 0 | Marbled murrelet (+) |
| Predation (herbivory) | – | Bear (+) |
| Mutualism | + | Fungi (+) |

2. Interactions:

| | B | Description |
|---|---|---|
| (a) | + | Both species benefit. |
| (b) | 0 | Species A benefits, no effect on species B. |
| (c) | + | A (host) is harmed, B (parasite) benefits. |
| (d) | + | A (prey) is harmed, B (predator) benefits. |
| (e) | – | Both species (competitors) are harmed. |

3. Predators kill and eat their prey. Parasites feed of their host but do not usually cause the death of the host.

4. Competition may require energy input from both parties (e.g. fighting for mates). More importantly, it reduces access to and availability of resources (e.g. food) for both parties (given that resources are limited).

5. Trees can only grow where sufficient light, nutrients, and water are available. Redwood density is therefore limited by the light available and the restrictions of root competition between neighboring plants. Trees growing too close together will shade each other and compete for nutrients and water, reducing each individual's growth. Trees will be spaced in a way that allows the growth requirements of all individuals to be met.

6. Redwoods depend on a mutualistic relationship with mycorrhizae and their growth is poor without them. Redwoods will only expand into regions where the requirements for mycorrhizal survival are also met.

7. The marbled murrelet nests in the old growth branches of redwoods, so the number of nests is dependent (at least in part) on the density and range of redwoods. Fewer old growth trees means fewer potential nesting sites. The decline of the marbled murrelet has been linked to the logging of old growth forests.

8. Bears strip the bark from trees for the sapwood after emerging from hibernation. Their feeding preferences are influenced by the density of tree species. If redwood density is too low, they will select other species (e.g. redwoods are preferred in Northern California, Douglas fir in the Pacific Northwest). In this way, they forage in the most energy-efficient manner (effectively 'prey switching' to feed on most abundant species).

9. The relationship could be called parasitism (although the oxpecker does not live on or in the host) or general exploitation as they feed on a host and cause damage without killing the host.

## KEY TERMS: Mix and Match (page 67)

Abiotic factors (U), Biome (I), Biosphere (E), Biotic factors (D), Commensalism (Z), Community (C), Consumer (G), Detritivore (V), Ecological niche (J), Ecology (F), Ecosystem (S), Environment (Q), Food chain (M), Food web (B), Habitat (N), Interspecific (K), Intraspecific (O), Microhabitat (R), Mutualism (AA), Parasitism (BB), Photosynthesis (X), Population (H), Producer (T), Realized niche (A), Respiration (W), Saprotroph (P), Species (L), Trophic level (Y)

## Ecosystem Stability (page 69)

1. C

2. **Keystone species** are pivotal to some important ecosystem function such as production of biomass or nutrient recycling. Because their role is disproportionately large, their removal has a disproportionate effect on ecosystem function.

3. (a) Sea otter: One of the favorite delicacies of the otter is the large sea urchin, which in turn feeds on kelp. Without sea otters there would be no kelp forests because the sea urchin would eat all the kelp. The diversity of the sea otter's diet of marine invertebrate herbivores and filter feeders reduces competition

© 2007-2013 BIOZONE International
ISBN: 978-1-927173-60-2
Photocopying Prohibited

between benthic grazers and supports greater diversity in those species.
(b) Beaver: When beavers build dams many species, some threatened or endangered, benefit. Beaver ponds produce food for fish and other animals, as well as creating habitat. Beaver activity is closely tied to the regeneration of quaking aspens. Beavers eat the bark of this (their favorite) tree and, by harvesting the trees, release buds for sucker growth and stand replacement.
(c) Gray wolf: Wolves, as a keystone predator, are an integral component of the ecosystems to which they belong. The wide range of habitats in which they thrive reflects their adaptability as a species. Their diet includes elk, caribou, moose, and deer as well as smaller prey. Wolves are sensitive to fluctuations in prey abundance, and the balance between wolves and their prey preserves the ecological balance between large herbivores and available forage.
(d) Quaking aspen: An aggressive pioneer species that frequently colonizes burned ground. The success of quaking aspen is attributed to its extensive root system, which sends up suckers to produce clones of the parent tree. The open canopies of aspen groves allow a rich and diverse under storey of shrubs, forbs, and grasses to feed and shelter a variety of wildlife. A large number of birds and browsing mammals are dependent on aspen stands for survival, especially through winter.

4. Humans historically kill off top carnivores when they enter a natural ecosystem and this drastically affects biodiversity and leads to ecological imbalances. For example, **wolves** were nearly hunted out of existence in the USA and Europe prior to the twentieth century. Following eradication of wolves in Yellowstone National Park in the early 1900s, elk numbers increased markedly, destroying vegetation and driving beavers and other animals from the damaged habitats. The pivotal role of top predators was determined only after ecological research during the last century and, as a consequence, wolves were reintroduced to Yellowstone and Idaho. The return of the wolves has resulted in a return of biodiversity as the ecological balance has been restored. Another similar case is the depletion of **sea otter** populations as a result of the fur trade from the mid 1700s to 1911. Removal of the otters resulted in a population explosion of sea urchins (on which the otters preyed) and destruction of the local kelp forests on which a large variety of smaller animals depended. As the sea otter populations recover following reintroductions to the natural range, populations of sea urchins are predicted to decline, allowing a recovery of marine plant biomass.

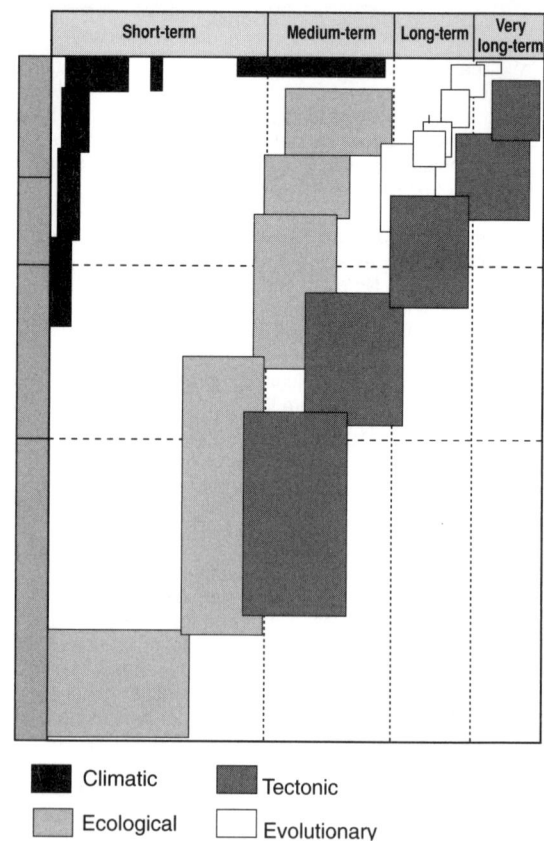

## Environmental Change (page 71)
1. (a) Ice ages or glacials
   (b) Drop in sea level by as much as 180 meters (590 ft), spreading of the ice caps from the poles (up to several kilometers thick).

2. Color as indicated in the diagram, top of next column.

3. The types of environmental change are linked because each has an effect on the others: the change in one will affect the others. The categories are simply convenient ways in which to examine the type, scale, and extent of environmental change.

## Nutrient Cycles (page 72)
1. (a) Bacteria are able to make conversions to and from elements and their ionic states. This gives plants and animals access to nutrients that they would otherwise not have (i.e. increases bioavailability).
   (b) Fungi decompose organic matter, returning nutrients to the soil where plants and bacteria can access them. They are also able to convert some nutrients into more readily accessible forms.
   (c) Plants are able make their own food and, when they die, add this to the soil in the form of nutrients that can be broken down and used by bacteria and fungi. They also provide browsing animals with nutrients when they are eaten.
   (d) Animals break down materials from plants, fungi, and bacteria and return them to the soil with their wastes and when they die allowing the nutrients in them to re-enter the cycle.

2. The rates of decomposition are very high in the higher temperatures of tropical forests. As a result, decaying matter is processed very quickly and very little remains in the soil. Much of the carbon and other nutrients are also locked up in biomass.

3. A macronutrient is one that is required in large amounts and often needs to be replaced on a regular basis by eating or drinking. They are often needed for growth and repair of the organism. Micronutrients (trace elements) are needed in much smaller amounts and may not need to be replaced very often. However they are often essential to the operation of biochemical pathways (e.g. as enzyme cofactors).

## The Carbon Cycle (page 73)
1. Arrows can be added for the points (a)-(d) as follows:
   (a) Dissolving of limestone by acid rain: Arrow from the limestone layer to atmospheric $CO_2$.
   (b) Release of carbon from the marine food chain: Arrows (labelled **respiration**) from marine organisms to atmospheric $CO_2$.
   (c) Mining and burning of coal: Arrow (**combustion**) from the coal seam to atmospheric carbon dioxide.
   (d) Burning of plant material: Arrow (**combustion**) from the trees and/or grassland to atmospheric $CO_2$.

2. (a) Coal: Plant material trapped under sediment in swampy conditions millions of years ago.
   (b) Oil: Marine plankton rapidly buried in fine sediment millions of years ago.
   (c) Limestone (also chalk = fine limestone): Shells of molluscs, skeletons of coral and other marine organisms with skeletons of calcium carbonate piled upon seabeds and compressed.

3. **Respiration** (stepwise oxidation of glucose) and **combustion** (rapid oxidation of organic substances accompanied by heat). Both involve the release of carbon dioxide.

4. (a)-(d) in any order: Atmosphere, coal, limestone, oil and natural gas.

5. (a) Photosynthesis
   (b) Respiration

6. Carbon would eventually be locked up in the bodies (remains) of dead organisms. Dead matter would not rot. Possible gradual loss of $CO_2$ from the atmosphere.

7. (a) Dung beetles: Bury the cow manure and the larvae feed on it. Burying the dung makes it available to decomposers in the soil. The beetle larvae reprocess the dung, using it as a food source. It therefore re-enters the trophic system.
   (b) Termites: Digest the cellulose in plant material, breaking it down and freeing up the carbon back into the ecosystem.
   (c) Fungi: Break down dead material, utilizing it as food and converting it into the fungal body. This makes it available to reenter the food chain.

8. Many insects play roles in the digestion of cellulose and animal wastes. Termites use bacteria in their gut to digest cellulose from woody trees. Beetles may eat animal waste or lay eggs on it so their grubs may use it as a food source

9. (a) Humans deplete fossil fuel reserves through mining (they provide a source of readily available energy).
   (b) The burning of fossil fuels increases the amount of carbon dioxide in the atmosphere, contributing to the rise in global temperatures. Burning also increases levels of air pollution.
   (c) Minimizing fossil fuels use through the use of alternative, environmentally clean sources of energy (solar energy, wind energy). Making sure that when fossil fuels are burnt, that combustion is as clean (complete) as possible, to minimize pollution.

## The Nitrogen Cycle (page 75)

1. (a)-(e) any of:
   - Decomposition or decay of dead organisms, to ammonia by decomposer bacteria (ammonification).
   - Nitrification of ammonium ions to nitrite by nitrifying bacteria such as *Nitrosomonas* ($NH_4^+ \rightarrow NO_2^-$)
   - Nitrification of nitrite to nitrate by nitrifying bacteria such as *Nitrobacter* ($NO_2^- \rightarrow NO_3^-$)
   - Denitrification of nitrate to nitrogen gas by anaerobic denitrifying bacteria such as *Pseudomonas* ($NO_3^- \rightarrow N_{2(g)}$)
   - Fixation of atmospheric nitrogen to nitrate by nitrogen fixing bacteria such as *Azotobacter* and *Rhizobium* ($N_2 \rightarrow NO_3^-$)
   - Fixation of atmospheric nitrogen to ammonia by nitrogen fixing cyanobacteria ($N_2 \rightarrow NH_3$)

2. (a) Oxidation of atmospheric nitrogen by lightning.
   (b) Nitrogen fixation (by bacteria).
   (c) Production of nitrogen fertilizer through the Haber process.

3. Denitrification

4. The atmosphere.

5. Nitrate.

6. Any one of: amino acids, proteins, chlorophyll.

7. Animals ingest food (plants or other animals) which contain nitrogen. Animals are heterotrophic.

8. Legumes are high in nitrogen. Ploughing them in replenishes soil nitrogen and reduces the need for additional nitrogen fertilizer when growing non-leguminous crops.

9. Human intervention in the nitrogen cycle by ((a)-(e) in any order):
   - Addition of nitrogen fertilizers to the land. This practice supplies inorganic nitrogen, as nitrate, for plant growth. However, any excess nitrogen, not absorbed by plants, may enter and pollute ground water and water bodies.
   - Industrial physical-chemical fixation of nitrogen (through the Haber process) combines hydrogen and nitrogen to ammonia, which can be used to manufacture inorganic nitrogen fertilizers. This is an industrial process, which requires high temperatures and pressures and uses a large amount of energy. The effects of applied inorganic nitrogen are outlined in (a) above.
   - Genetic modification of crop plants so that they can fix nitrogen. The effect of this is to increase the range of crop plants capable of growing on nitrogen deficient soils. Potentially, this could make a beneficial contribution to soil fertility
   - Large-scale, assisted composting produces nitrogen rich organic fertilizer which has the effect of improving soil fertility and structure. This has beneficial effects in reducing the amount of inorganic nitrogen fertilizer that must be applied.
   - Burning and harvesting removes nitrogen from the land and releases nitrogen oxides into the atmosphere.
   - Discharge of effluent (e.g. animal waste) into waterways enriches water bodies and leads to localized pollution and eutrophication.
   - Irrigation can accelerate loss of nitrate from the soil by increasing the rate at which nitrates are washed out of the soil into ground water.

## Nitrogen Pollution (page 77)

1. (a) NO contributes to the formation of low level ozone which is a constituent of photochemical smog.
   (b) $N_2O$ depletes ozone in the upper atmosphere.
   (c) $NO_2$ is a toxic inhalant. It also contributes to the formation of nitric acid in the atmosphere and therefore acid rain.
   (d) $NO_3^-$ is a water pollutant. It causes eutrophication, the accelerated growth of algae in waterways, and can cause severe health problems if drinking water contains significant amounts.

2. NO persists in the atmosphere both causing and being released by cyclic chemical reactions. NO reacts with oxygen to form toxic $NO_2$, which reacts with water to form $HNO_3$ (acid rain) and $HNO_2$. The $HNO_2$ decomposes, releasing NO to react again. NO will continue this cycling until it reacts with a chemical that removes it from the cycle.

3. Even after nitrogen fertilizers are not used, there is a large nitrogen load in soil and groundwater. Groundwater may take many years to move from its point of origin to its point of exit. Nitrate fertilizers that leach into groundwater now will move with this ground water and exit into waterways many years afterwards. In some cases, the lag may be up to fifty years.

4. (a) 1860: Reactive N deposition in the ocean = 156.5 million tonnes. Release of unreactive nitrogen = 301 million tonnes.
   1995: reactive N deposition in the ocean = 202 million tonnes. Release of unreactive nitrogen = 322 million tonnes. This is an increase of 45.5 million tonnes of reactive N deposition but an increase of only 21 million tonnes of unreactive nitrogen released Result: twice as much reactive nitrogen has been added to the ocean than has been released as unreactive nitrogen.
   (b) Algal blooms are becoming more common in the oceans as nitrate levels slowly rise. Many of these blooms are from algae that contain small amounts of toxins. These can be concentrated by filter feeders such as mussels and if eaten can cause poisoning.

5. (a) Nitrates are highly soluble in water and a lot is rapidly washed away or leached from the soil and not incorporated into plant tissues. Nitrates are also broken down by bacteria and returned to the air. Some nitrates will accumulate in the soil over time but not be accessible to plants. All these factors contribute to "lost" nitrogen.
   (b) Nitrogen losses could be minimized by fertilizer application at appropriate times and rates, and by sensible irrigation practices. Using slow release fertilizers in times of frequent rain also slows down the rate at which nitrates are lost into groundwater.

## The Hydrological Cycle (page 79)
1. (a) Surface runoff
   (b) Ground-water flow

2. (a)-(c) any of the following, in any order:
   - Humans withdraw water from ground-water storage, rivers, and lakes. It may be used to supply domestic or personal use, or for irrigation. Consequently it may become depleted in specific areas or its normal destination altered.
   - Humans divert water and alter natural flows through damming and controlled flows. This alters the normal balance of seasonal water movements.
   - Humans may use water courses or water bodies for disposal of waste, polluting it and making it unsuitable for other organisms.
   - Humans clear vegetation, reducing the amount of water re-entering the atmosphere and being returned to the land via precipitation.

3. The oceans

4. In descending order of magnitude: Snow and ice (in ice sheets and glaciers), ground-water, lakes, soil, atmosphere, rivers.

5. Plants lose a vast amount of water through transpiration. This is returned to the atmosphere where it condenses and then precipitates back to the land.

## The Phosphorus Cycle (page 80)
1. Arrows from guano deposits and rock phosphate (**mining/removal**) to: "Dissolved phosphates available to plants ($PO_4^{3-}$)".

2. (a) 
   - Decomposers release organic molecules (ATP, DNA, etc.) that can be further broken down.
   - Phosphatizing bacteria release soluble phosphate.
   
   (b) Any two of: DNA, ATP, phospholipids.
   
   (c) 
   - Rock phosphate: Much phosphate is washed into the ocean where it builds up in phosphate-rich rocks made from marine sediments.
   - Bone deposits: Remains of dead marine vertebrates washed down rivers into lakes and into the sea.
   - Guano deposits: The droppings of birds (especially fish-eating birds) accumulated at nesting colonies. Cave dwelling bats also produce guano deposits.

3. Geological uplift and weathering (erosion).

4. Any one of: Phosphorus, unlike carbon, has **no** (significant) **atmospheric component**. **Cycling** of phosphorus **is slow** and tends to be localized.

## The Sulfur Cycle (page 81)
1. (a) As $SO_2$ and sulfate from volcanic activity, hot springs, and biogenic activity.
   (b) As $H_2S$ gas arising from reduction of sulfur in soil and sediments.

2. (a)-(b) any two of:
   - As $SO_2$ from combustion of fossil fuels.
   - As sulfur oxides from smelting.
   - As sulfur oxides from refining petroleum.

3. (a) Decomposition.
   (b) Microbial oxidation of $H_2S$ (via elemental sulfur) to sulfates (anaerobic and aerobic).
   (c) Acid precipitation.

4. (a) Minor but essential component of proteins.
   (b) Ecological role in determining the acidity of precipitation, surface water, and soil.

## Primary Succession (page 82)
1. Glacial retreat, exposed slip, new volcanic island.

2. (a) Lichens and bryophytes (mosses and liverworts), as well as some hardy annual herb species, are often the first to colonize bare ground.
   (b) Chemically and physically erode the rock (making the beginnings of a soil) and add nutrients by decay.

3. Climax communities tend to have greater biodiversity and a more complex trophic structure than early successional communities. A greater diversity of community interactions buffers the system against disturbances because there are many more organisms with different ecological roles able to compensate for losses from the system.

## Secondary Succession (page 83)
1. Primary succession refers to the colonization of regions where there is no preexisting community (e.g. rocky slope, exposed slip, new volcanic island). Changes in the community occur in stages until a climax community is reached. Secondary succession follows the interruption of an established climax community (e.g. logging, pasture reverting to bush).

2. Secondary succession proceeds more rapidly than is the case with primary succession because although the land is cleared, there is minimal or no loss of soil or seed stores. Many plants may still be able to grow despite the disturbance and the climax community will reestablish faster because nutrients are already available and seeds already laid down.

3. Any one example: Events that result in secondary succession include forest fires, minor landslides, storm or cyclonic damage, the fall of a canopy tree, flooding, or human induced land clearance.

4. (a) Selective logging causes gaps in the canopy with the loss of the large trees. This results in a chance for new seedlings to gain more light and grow up. These seedlings may be the same or different from the species that was/were removed. Forest composition is changed by becoming temporarily more open, with a loss of the dominant large trees.
   (b) Selective logging is considered (by some) to be preferable to clear felling because there is still a favorable environment in which seedlings can mature. Canopy gaps are created naturally by windfalls; selective logging is said to mimic this, so the gaps created may be beneficial for forest diversity and regeneration. Clear felling, in contrast, removes all forest from the area leaving an environment unfavorable for the regeneration of existing forest species but favorable for weeds.

5. (a) A deflected succession refers to a succession that is deflected from its natural course by human intervention. The plagioclimax that develops is different from the one that would have developed if the intervention had not occurred.
   (b) A deflected succession keeps the managed habitat as a modified environment in a certain state. Many human-modified landscapes (e.g. farm lands) are managed (by burning, grazing, mowing) with the express purpose of preventing the establishment of a natural climax community. These communities are distinct from those that would naturally develop if the land were left alone.

## A Case Study in Primary Succession: Surtsey Island (page 85)
1. Surtsey was ideal as a study site for primary succession because it was an entirely new island, devoid of any soil, and was isolated from nearby influences (such as already established vegetation or urban settlements) that could accelerate the succession process.

2. Early colonizations were primarily influenced by the island's location to the south of Iceland, so the northern shores are closest to a land mass. Later colonizations were influenced by the establishment of a gull colony at the southern end of the colony. The gulls would transport seeds and contribute to soil fertility.

3. (a) 1985.
   (b) Transported by birds.

© 2007-2013 BIOZONE International
ISBN: 978-1-927173-60-2
Photocopying Prohibited

(c) 1985. This coincides with the establishment of the gull colony as the gulls were instrumental in dispersing seeds.

---

### KEY TERMS: Mix and Match (page 86)
Biogeochemical cycling (F), Carbon cycle (G), Climax community (U), Cultural eutrophication (I), Decomposer (D), Eutrophication (J), Hydrological cycle (K), Keystone species (B), Nutrient cycling (E), Nitrifying bacteria (L), Nitrogen cycle (S), Nitrogen fixing bacteria (N), Phosphorus cycle (O), Pioneer species (P) Pioneer community (Q), Primary succession (A), Resilience (R), Secondary succession (C), Sere (M), Sulfur cycle (T), Stability (H)

---

### The Rise and Fall of Human Populations (page 88)
1. The trend is of continual (close to exponential) growth in the human population.

2. The human population has grown because of increased crop yields and better medical treatment. This has led to a higher standard of living, better nutrition, and a lower mortality rate.

3. Local resources at Tikal and Easter Island were both used to support population growth. When these resources were over-exploited, the populations crashed and the lands were abandoned. The early history of these examples tends to mirror what appears to be currently happening on a global scale. This helps us understand the effects of depleted resources on populations whose livelihoods are based on those resources and what can be done to prevent similar events happening again.

---

### Features of Populations (page 89)
1. (a) Any one of the following:
   **Population growth rate**: If this increases (or decreases) from one time interval to the next, it indicates that the population is probably also increasing (or decreasing). **Note**: The intrinsic rate of population increase ($r_{max}$) should be distinguished from population growth rates that account for the increasing number of individuals in the population ($rN$). The intrinsic rate of population increase is characteristic for each species but $rN$ can increase rapidly as more and more individuals add to the population increase (giving an exponential curve). Population growth rates account for birth and death rates but do not usually account for losses and gains through migration, which are usually assumed to be equal.
   **Total abundance**: If this increases (or decreases) from one time interval to the next, it indicates that the population is also increasing (or decreasing).
   **Mortality**: If this is increasing from one time interval to the next, it indicates that the population may be decreasing. You must also account for other sources of population change (e.g. birth rates, migration).
   **Birth rate & population fertility**: If these increase from one time interval to the next, they indicate that the population may be increasing. You must also account for other sources of population change (e.g. mortality and migration).
   **Age structure**: A population dominated by young individuals is usually increasing. A population dominated by old (especially post-reproductive) individuals is usually decreasing.
   (b) Any one of the following:
   **Distribution**: A very clumped distribution may indicate that only some parts of the environment are suitable for supporting individuals.
   **Population growth and birth rates**: If these are low or declining it may indicate an inability of the environment to support the population density.
   **Mortality**: If this is very high or increasing it may indicate an inability of the environment to support the present population density.

2. (a) **Measurable attributes**: Density, distribution, total abundance, sex ratios, migration (sometimes difficult). In some cases, depending on the organism, also age structure and population fertility.
   (b) **Calculated attributes**: Population growth rate, natality (birth rate) and mortality (death rate).

3. (a) Population sampling of an endangered species allows us to determine (any of): How fast a population is growing (if at all); what the age and sex structure of the population is (i.e. is it dominated by young or very old, non-reproductive, individuals); population abundance, density and distribution in different areas (giving an idea of habitat preference and suitability); sources of mortality (predation, disease, starvation etc.); population fertility (are the reproductive individuals in good shape or not). This type of information allows more informed decisions to be made about the current status of the population and how best to manage it (through habitat restoration or captive breeding for example).
   (b) Population sampling of a managed fish species allows us to determine the population growth rate. This is critical to establishing the level of fishing that can be supported by the population (the sustainable harvest) without irreversible population decline. The growth rate is calculated taking into account population abundance, and birth and death rates. Sustainable harvest can be built into the equation as one of the (controllable) sources of mortality.

---

### Density and Distribution (page 90)
1. (a) Resources such as food and shelter are not usually spread through the environment in an even manner. Organisms will clump around these resources.
   (b) Some organisms group together for protection from the physical environment or from predators. They may also group together for mating and reproduction. Clumped distribution may also result from the method of dispersal (especially in plants where they may use two methods; seeds which are well dispersed and vegetative means which remain close to the parent).

2. Territorial behavior whereby an area occupied by an animal, or by a pair or group of animals is defended against intruders.

3. Resources in the environment are limited but are distributed uniformly.

4. (a) **Clumped**: Many marine gastropods, colonial birds (seasonally), many mammals that exhibit grouping/herd behavior, schooling fish, colonial insects, many other invertebrates such as coral, some plants with limited dispersal.
   (b) **Random**: Weed plants with effective dispersal method, shellfish on sand or mud substrate.
   (c) **Uniform**: Territorial organisms, monoculture plantings (e.g. crops, timber plantations).

---

### Population Regulation (page 91)
1. **Density dependent factors**, such as disease, parasitic infestation, competition, and predation have an increasing effect on population growth as the density of the population increases; their effects are exacerbated at high population densities because they are driven in part by the number of organisms present. **Density independent factors**, such as flood, fire, and drought have a controlling effect on population size and growth that is independent of the population density. The severity of the impact on the population is not correlated with population density.

2. When population density is low, individuals are well spaced apart. This can reduce stress between individuals (improving the resistance to diseases) as well as making the transmission of the disease more difficult. Disease is more easily spread between individuals in a crowded population. Epidemics of infectious disease are more likely to occur.

3. (a) Density dependent factor: Predation (e.g. by ladybird beetles), competition with other aphids for position on the best part of the plant to feed.
   (b) Density independent factor: Temperature. A drop in temperature in autumn (fall) in cooler climates causes the population to crash.

## Population Growth (page 92)

1. (a) Mortality: Number of individuals dying per unit time (death rate).
   (b) Natality: Number of individuals born per unit time (birth rate).
   (c) Net migration rate: Net change in population size per unit time due to immigration and emigration.

2. (b) Declining population:   B + I < D + E
   (c) Increasing population:   B + I > D + E

3. Rate of change for USA: + 1.0
   Rate of change for Mexico:    + 3.3

4. (a) Birth rate = 14 births ÷ 100 total number of individuals x 100 % = 14% per year
   (b) Net migration rate = 2% per year
   (c) Death rate = 20% per year
   (d) Rate of population change: birth rate – death rate + net migration rate =  14 – 20 + 2 = -4% per year
   (e) The population is declining.

## Survivorship Curves (page 93)

1. (a) C Type III      (b) A Type I      (c) E Oyster

2. They produce vast quantities of eggs/offspring.

3. This statement is realistic because particular patterns of survivorship are hypothetical and only provide a tool for categorizing the characteristics of populations. The survivorship curves of many species may show a mix of Type I, II, and III characteristics depending on the life cycle stage. Many extrinsic factors may also influence mortality and it is not necessarily possible to accurately predict a steady-state pattern of survivorship for any one species.

## Life Expectancy and Survivorship (page 94)

1. (a) As GDP increase so does life expectancy. When plotted on a graph a curved line is formed.
   (b) Wealthier nations generally have well developed health care and social systems which provide better health care to new born babies and infant children.
   (c) Factors that lower life expectancy may be: war, social unrest, poverty, and poor health resources.

2. Life expectancy is a measure of the number of years a person is expected to live for at any give age. It is based on the probability of living to the next year of life. Different stages of life leave people more vulnerable than others, so life expectancy changes as one moves through these areas of life. A new born baby is highly vulnerable and has a lower life expectancy than a one year old. An 80 year old is also vulnerable and has a low life expectancy. However a 90 year old has a relatively high life expectancy as people who reach this age tend to then live a number of years longer.

3. (a) 78       (b) 70       (c) 63

## Population Growth Curves (page 95)

1. As population numbers increase, the resistance of the environment (to further population increase) increases. This constrains the population to keep to a size that the environment can support at any one time.

2. Environmental resistance refers to all the limiting factors that together act to prevent further population increase (achievement of intrinsic rate of population increase, $r_{max}$).

3. (a) The maximum population size (of a species) that can be supported by the environment.
   (b) Carrying capacity limits population growth because as the population size increases, population growth slows (when N = K population growth stops). **Note:** For those interested in extension in this area, the effect of K on population growth is defined by the mathematical expression of logistic growth. This is covered in many, more advanced, biology texts.

4. (a) A new introduction increases exponentially (or nearly so) in a new area because its niche in that environment is unexploited up to that point. Resources (food, space, shelter etc.) are plentiful and readily available. The population rapidly increases, then slows as the population encounters environmental resistance.
   (b) Population numbers would fluctuate around some relatively stable population size that equates to what the environment can support (the carrying capacity).

5. Introduced grazing species can lower the carrying capacity of environments by reducing the ability of the environment to recover from the impacts of grazing.
   **Note:** High population numbers and high stocking levels (e.g. sheep in Australia, cattle in sub-Saharan Africa) lead to overgrazing and trampling of the soil. Soil is lost through erosion and desirable plant species are then replaced by (weed) species that can survive the grazing pressure. Native consumers tend not to overexploit the environment in this way because they have different patterns of resource use and population growth.

## r and K Selection (page 96)

1. r refers to the maximum reproductive potential of an organism and r-selected species are those with a high intrinsic capacity for population increase. K refers to the carrying capacity of the environment and K-selected species exist near this point of equilibrium with the environment.

2. r-selected species are opportunists because they are poor competitors and must continually invade new areas in order to gain the advantage of their high reproductive potential. Examples include algae, bacteria, rodents, many insects and most annual plants.

3. K-selected species are also called competitor species because they are challenged in competitive environments to use available resources more efficiently and thereby compensate for their lower reproductive potential. Examples include most larger mammals, birds of prey, and large, long-lived plants.

4. Many K-selected species are vulnerable to extinction because their lower reproductive potential makes it difficult for them to recover stable, viable populations following loss (of numbers or suitable environments).

## Population Age Structure (page 97)

1. (a) 3:1
   (b) Other factors besides changes in age structure can affect population growth, e.g. sex ratios, population fertility, and migration.

2. Over a short duration, the large cohort of prereproductive individuals will reach reproductive age and the population will continue to grow. Even if the rate of population growth continues to slow it will take several generations before the 'bulge' of reproductive individuals moves into the post-reproductive phase.

3. (a) Mortality
   (b) Higher proportion of smaller/younger fish.

4. (a) 3 years       (b) 5 years       (c) 8 years

5. (a) Gray face: It has palms of all sizes and therefore all ages are represented.
   (b) Golf course: No young plants are represented.

6. The population will age, with the established palms growing taller, and no new palms becoming established. Eventually these older palms will die with no replacement (unless there is a planting program).

7. Not all organisms (e.g. plants, fish) grow at the same rate. Size may depend on the quality and quantity of food supply. Some seasons may produce more growth than others.

8. If the age structure in the short-medium term shows a trend to smaller/younger age classes, then harvesting pressure

© 2007-2013 BIOZONE International
ISBN: 978-1-927173-60-2
Photocopying Prohibited

is too severe. If this continues, there will be few individuals of reproductive age and, consequently, a decline in the harvestable stock (population size).

## World Population Growth (page 99)

1. (a) Africa
   (b) Poor education on family planning. Entrenched cultural practices (large families are desirable).

2. The increased mechanization of agriculture and move away from subsistence farming has reduced the need for large numbers of rural workers and encouraged the move of people to cities.

3. (a) Positive effects of urbanization include:
   - opportunity for better jobs.
   - improved housing.
   - larger workforce concentrated in one area.
   - increased accessibility of community resources and services (e.g. education, healthcare).
   - greater social integration (greater understanding and acceptance of cultural differences).
   (b) Negative effects of urbanization include:
   - problems with adequate provision of food, water, sanitation, housing, jobs, and basic services (such as health care and education).
   - the rapid and unplanned growth of slum or squatter areas that develop on the fringes of cities. These areas often grow rapidly in response to immigrant influxes.
   - increased levels of pollution as a direct or indirect result of increased population pressure.
   - rapid spread of diseases through high density and/or improperly serviced areas.
   - increased crime, particularly in high density, impoverished parts of cities.

## Human Demography (page 101)

1. Diagram C corresponds to stage one of the DTM. The wide base of the pyramid indicates a large number of under 15s (pre-reproductive) but the sides of the pyramid are steep and not many live to old age, so death rates are high. The combination of high birth rates (many children) and high death rates means the population numbers are largely stationary.

2. In less economically developed nations (LEDN), children are required to contribute to the household economy by helping with work and raising younger siblings. In more economically developed nations (MEDN), the household income is more likely to come from outside industry (i.e. wage and salary earning rather than farming, herding animals, or making crafts to sell for example) so extra children become a cost to the household rather than contributors in their own right. This is especially the case because the costs of child rearing (education, clothing, transport etc) are higher in MEDN.

3. (a) and (b) see tables below.

| Age | Males Pre-1950 | | Females Pre-1950 | |
|---|---|---|---|---|
| | No. of deaths | Survivorship | No. of deaths | Survivorship |
| 0-9 | 5 | 30 | 7 | 30 |
| 10-19 | 1 | 25 | 2 | 23 |
| 20-29 | 4 | 24 | 2 | 21 |
| 30-39 | 2 | 20 | 1 | 19 |
| 40-49 | 4 | 18 | 3 | 18 |
| 50-59 | 2 | 14 | 2 | 15 |
| 60-69 | 3 | 12 | 7 | 13 |
| 70-79 | 7 | 9 | 4 | 6 |
| 80-89 | 2 | 2 | 2 | 2 |
| 90-99 | 0 | 0 | 0 | 0 |
| Total | 30 | | 30 | |

| Age | Males Post 1950 | | Females Post-1950 | |
|---|---|---|---|---|
| | No. of deaths | Survivorship | No. of deaths | Survivorship |
| 0-9 | 1 | 30 | 0 | 30 |
| 10-19 | 1 | 29 | 0 | 30 |
| 20-29 | 0 | 28 | 0 | 30 |
| 30-39 | 2 | 28 | 1 | 30 |
| 40-49 | 3 | 26 | 2 | 29 |
| 50-59 | 1 | 23 | 4 | 27 |
| 60-69 | 5 | 22 | 4 | 23 |
| 70-79 | 7 | 17 | 4 | 19 |
| 80-89 | 8 | 10 | 10 | 15 |
| 90-99 | 2 | 2 | 5 | 5 |
| Total | 30 | | 30 | |

4. (a)

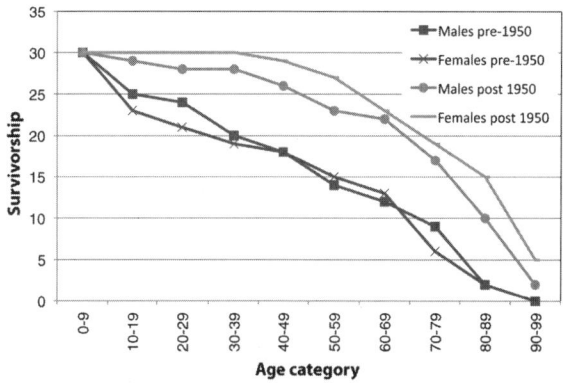

Survivorship for males and females in the US pre-1950 and post 1950

(b) Survivorship for both males and females has improved since 1950.
(c) The pre-1950s had two world wars in which many young men were killed. Diseases such as TB and polio also killed many young people. In contrast, many medical advancements have been made in the decades since 1950, these have increased life expectancy. There has also been less global conflict (fewer men dying in war).

## Human Sustainability (page 103)

1. Many developing countries still lack easy access to family planning advice and contraception. Governments may not have programs in place to educate the public about family planning. Often culture promotes larger families as desirable. Sometimes many children are seen as desirable because they can contribute to the household income (e.g. working on family land).

2. Public education and incentives that get people to make a change in their life style are usually more successful than ones that do not. The success of China's family planning campaign hinged on three main points: 1) public education, 2) Incentives backed by the government, 3) China's semi-dictatorial government was able to impose population controls consistently and without fear of a public backlash. India was not able to impose such stringent controls because of: 1) A lack of government support and incentives, 2) Poor planning, 3) No compulsion for families to change their behaviors.

3. (a) Reducing population size allows countries to better control the use of resources and be more reliant on domestic resources. If a population growth rate is too high a country risks using up its resources and not being able to keep up with infrastructure development. This can lead to unemployment and poor living conditions.
   (b) A rapidly declining population can lead to a reduced labor force, a reduction in tax revenue. This reduces a countries ability to take advantage of new resources

and opportunities. Infrastructure development may fall behind.

### Humans and Resources (page 104)
1. The modern industrialized agricultural techniques on which global food production is currently based are heavily reliant on plentiful supplies of water and fossil fuels, yet supplies of both are dwindling. World-wide aquifers are being depleted at rates far exceeding replenishment and coming decades, especially following peak oil (when oil supplies peak and begins to decline), will see sharp increases in oil prices. Likely effects on agriculture include:
   - a rise in the costs of food production and therefore on the costs of food to consumers.
   - conversion of food crops over to fuel production and direct competition between fuel producers and food processors for supplies of wheat, corn, soybean, sugarcane, and other key crops.
   - water short countries diverting water from irrigation and importing grain to feed their populations (with a consequent decline in food production overall).
   - more countries with grain deficits.
   - increased farming of more marginal land.
   - greater support for more sustainable farming practices that reduce reliance on water and fossil fuels and concomitant economic discouragement of unsustainable practices (through levies on water and fuel use).

2. (a) and (b)
   - Water is used in industry, agriculture (irrigation and feed), and domestic (washing, sewage treatment). Impacts of its use include reduction of water flow in waterways, affecting environments downstream.
   - Coal is used for industrial and domestic heating and for coking in industry. Its use produces vast quantities of carbon dioxide, soot and other hazardous gases.
   - Timber is used for building and domestic heating and cooking. Its use can cause deforestation and erosion from loss of trees. Air pollution may be cause from burning wood or soil particles blown away from newly cleared land.
   - Fish are used for eating, fishmeal for aquaculture and fish products (e.g. oil) as health supplements. Overfising has seriously reduced fish stocks in many species.

### KEY TERMS: Mix and Match (page 105)
Carrying capacity (F), Community (E), Competition (D), Demographics (M), Density (W), Distribution (A), Emigration (N), Environmental resistance (K), Exponential growth (C), Immigration (J), K-selection (H), Limiting factors (P), Mortality (Q), Natality (I), Population size (R), r-selected species (G), Sigmoidal growth (O), Survivorship curve (B), Urbanization (S)

### Sampling Populations (page 107)
1. We sample populations in order to gain information about their abundance and composition. Sampling is necessary because, in most cases, populations are too large to examine in total.

2. (a) Random or systematic quadrat sampling.
   (b) Random or systematic point sampling (using a net or trap).
   (c) Line transect with point sampling (from low to high altitude). If time for sampling and analysis is not constrained, a belt transect using quadrats at regular intervals would provide the most information.

3. Information about the physical environment helps to explain species distributions. Certain species are usually associated with a particular suite of physical factors (e.g. preferred exposure, temperature, humidity etc) and if these are measured and known, more information on community patterns can be gathered.

4. As the vertical distance up the trunk increases, light intensity and temperature increase and humidity declines. With these changes in physical conditions there is consequent change in the vegetation from a diverse community of shade and moisture adapted moss species, to a community comprising just one species of (hardier) tree moss and various species of lichens, i.e. species more tolerant of the microclimate of lower humidity and higher light and temperature.

### Quadrat Sampling (page 109)
1. Mean number of centipedes captured per quadrat:
   Total number centipedes ÷ total number quadrats
   = 30 individuals ÷ 37 quadrats
   = 0.811 centipedes per quadrat

2. Number per quadrat ÷ area of each quadrat
   $0.811 ÷ 0.08 = 10.1$ centipedes per m$^2$

3. Clumped or random distribution.

4. Presence of suitable microhabitats for cover (e.g. logs, stones, leaf litter) may be scattered.

### Quadrat-Based Estimates (page 110)
1. Species abundance in plant communities can be determined by using quadrats and transects, and abundance scales are often appropriate. Methods for sampling animal communities are more diverse, and density is a more common measure of abundance.

2. Size: Quadrats must be large enough to be representative and yet small enough to minimize the sampling effort.

3. Habitat heterogeneity: Diverse habitats require more samples to be representative because they are not homogeneous.

4. (a) and (b) any two of:
   - The values assigned to species on the abundance scale are subjective and may not be consistent between users.
   - An abundance scale may miss rarer species and overestimate conspicuous ones.
   - The scale may be inappropriate for use in some habitats.
   - The semi-quantitative values assigned to the abundance categories cover a range so results will lack precision.

### Sampling a Leaf Litter Population (page 111)
It must be emphasized that the actual results for this practical are not particularly important. What is of value, is learning the value and limitations of this method, preferably before students are asked to carry it out in a field situation. The actual results will vary, depending upon the group's agreed upon criteria for inclusion of organisms in a given quadrat (i.e. when and how an organism is counted when it is partly inside a quadrat).

**NOTE:** Some leaves are almost completely obscured by invertebrates or have other leaves on top of them.

6. Typical results for samples used are:

| | Direct count | A | B | C | D |
|---|---|---|---|---|---|
| Woodlouse: | 89 | 9.5 | 5 | 13 | 14.5 |
| Centipede: | 3 | 0 | 0 | 1 | 1 |
| False scorpion: | 3 | 1 | 0 | 1 | 0 |
| Springtail: | 6 | 0 | 3 | 0 | 0 |
| Leaf: | 168 | 29 | 20.5 | 24.5 | 26.5 |

Results will vary depending on counting criteria.

7. Typical results for calculated density are:

| | Direct count | A | B | C | D |
|---|---|---|---|---|---|
| Woodlouse: | 2747 | 1759 | 926 | 2407 | 2685 |
| Centipede: | 93 | 0 | 0 | 185 | 185 |
| False scorpion: | 93 | 185 | 0 | 185 | 0 |
| Springtail: | 185 | 0 | 556 | 0 | 0 |
| Leaf: | 5185 | 5370 | 3796 | 4537 | 4907 |

Area of 6 quadrats = $(0.03 \times 0.03) \times 6 = 0.0054$ m$^2$
Area of total sample area = $0.18 \times 0.18 = 0.0324$ m$^2$

8. (a) Problems with sampling moving organisms: Once the quadrats have been laid, the animals moving from one quadrat to another risk being counted twice.
   Solutions: The quadrat could involve the placement of physical barriers between each quadrat (what about

© 2007-2013 BIOZONE International
ISBN: 978-1-927173-60-2
Photocopying Prohibited

the invertebrates directly underneath). Possibility of exposing the entire area and photographing it for later analysis.
(b) Exemplar data given in the tables above. Students should be aware of the significance of extrapolating data from a small sample. The inclusion or exclusion of single individuals may have a large effect on the calculated density, particularly where species occur in low numbers.

**Extension**: Groups could combine their data to see if they get a more representative sample.

---

## Transect Sampling (page 113)

1. (a) With transects of any length (10 m or more), sampling (and sample analysis) using this method is very time consuming and labor intensive.
   (b) Line transects, although quicker than belt transects, may not be representative of the community. There may be many species which are present but which do not touch the line and are not recorded.
   (c) Belt transects use a wider strip along the study area and there is much less chance that a species will not be recorded.
   (d) It is not appropriate to use transects in situations involving highly mobile organisms.

2. To test whether or not the transect sampling interval was sufficient to accurately sample the community, the sampling interval could be decreased (e.g. from a sampling interval of every 1.5 m to an interval of every 0.25 m). If no more species are detected and the trends along the transect remain the same, then the sampling interval was adequate.

3. Distribution of *Littorina* species along a rocky shore. This figure has been laterally compressed to fit this format.

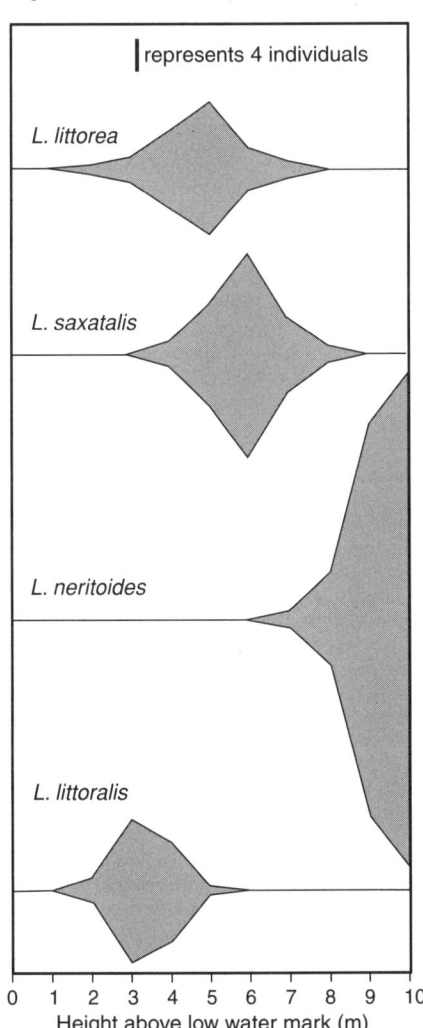

## Mark and Recapture Sampling (page 115)

1. Results will vary from group to group for this practical. The actual results are not important, but it should serve as a useful vehicle for discussion of such things as sample size, variation in results between groups, and whether the method is a reliable way to estimate the size of a larger unknown group. Discussion could center around what factors could be altered to make it a more reliable method (e.g. larger sample size, degree of mixing, increasing number of samples taken).

2. **Trout in Norwegian lake**:
   Size of 1st sample: 109
   Size of 2nd sample: 177
   No. marked in 2nd sample: 57
   Est. total population: 109 x 177 ÷ 57 = 338.5

3. (a) Some marked animals may die or disperse out of the sampling area.
   (b) Not enough time for thorough mixing of marked and unmarked animals.

4. (a) and (b) any two of:
   - Marking does not affect their survival.
   - Marked & unmarked animals are captured randomly.
   - Marks are not lost.
   - The animals are not territorial (must mix back into the population after release).

5. (a) Any animal that cannot move or is highly territorial (e.g. barnacle, tube worm, many mammals).
   (b) animals which cannot move, or stay within a territory will not mix with unmarked portion of the population. Recapture at the same location would only sample the same animals again.

6. (a)-(c) in any order:
   - Banding: leg bands of different color on birds.
   - Tags: crayfish shell, fish skin, mammal ears.
   - Paint/dye used to paint markings in shell/fur.

7. The scientists obtain information on fish growth to establish the relationship between age and growth. This will help manage the population to prevent over-fishing. Tracking also helps to map breeding grounds and migrations so that fish can be protected at critical times in their life histories. In addition to these data, researchers will find out more about the general biology of the cod (e.g. data on feeding), which will help in the long term management and recovery of the fish stock.

---

## Sampling Animal Populations (page 117)

1. The Tullgren funnel provides the best quantitative invertebrate sampling method. It is the only method that ensures that all of the invertebrates contained within a known area are removed, identified and counted to give a true reflection of population composition.

2. Pitfall traps rely on being placed in an area where the organisms are active. The traps take no account of clumped distributions or microhabitat preferences, and may overestimate densities in some areas and underestimate them in others.

3. (a) A large mesh size may fail to capture some smaller species or life stages. A fine mesh is apt to clog, and this reduces filtering efficiency so that much of the sampled volume is pushed out of the net instead of passing through it.
   (b) Mesh size should be fine enough to capture most or all of the species in which you are interested and it should be coarse enough to filter efficiently.

---

## Indirect Sampling (page 118)

1. (a) The Frog Census Datasheet would be able to gather information on the distribution of various frog species, population size (from the number of frogs heard calling), and their habitat quality (from the location, habitat assessment, water quality, and weather data).
   (b) Linking species abundance to habitat quality enables biologists to speculate on the cause of a species

demise and devise a plan of action to try to save a population under threat of habitat destruction.

2. Other indirect methods of population sampling include (any one of the following): Number of fecal pellets, frequency of calls, pelt records, number of artefacts (e.g. burrows, nests, pupal cases), gut contents of predators, questionnaires for hunters/recreational fishers, feeding capacity (bait taken) before and after poisoning, frequency of carcasses found on the road. Advantages include the estimate of populations of organisms that are normally elusive, easily disturbed or widely dispersed.
Disadvantages of using indirect sampling methods are that the measures of abundance are less reliable than direct sampling methods.

### Monitoring Water Quality (page 119)

1. Each of these water quality measurements must be made in the field as these physical factors may immediately change if a sample is removed for later analysis (e.g. water sample will change temperature in the container, oxygen may be gained or lost, suspended matter may settle to the bottom thereby changing clarity).

2. Many land-based activities result in intentional or accidental discharges into waterways which may affect water quality. Surface runoff during rain washes chemicals, silt, and organic matter into waterways.

3. Species diversity index can be used to provide any of the following information:
   - Comparisons of similar ecosystems which have been subjected to (beneficial or detrimental) human influence (e.g. restoration or pollution).
   - Assessment of the same ecosystem before and after some event (fire, flood, pollution, environmental restoration).
   - Assessment of the same ecosystem along some environmental gradient (e.g. distance from a point source of pollution).
   - Assessment of the biodiversity value of an area for the purposes of management or preservation (tends to be a political lobbying point).

4. Indicator species are species which are used to assess the health of an ecosystem. They can be used to detect pollution in a stream. The diversity index, based on the diversity of macroinvertebrates found in a stream community, will be low in a steam that is polluted, even though abundance may be high.

5. A BOD test could measure a series of BODs along a stretch of the water system. The point where the BODs suddenly change from low to high will indicate the entry point of the pollutant into the water systems.

### Radio-tracking (page 121)

1. (a) and (b) any two of:
   - Recording the extent of their home range and the types of habitats they are using.
   - Recording dispersal after release into a new area (e.g. offshore island).
   - Recording the areas they are using (e.g. for feeding) which may be outside of the protected area they are living in (bats are a good example of this).
   - Recording activity in relation to weather conditions and time of day (for some species, the accuracy of a population census depends on recording the animals at appropriate times).
   - Monitoring migratory routes so that the species can be protected during their migration.
   - Monitoring physiological state and learning more about the species biology.

2. In order to control pests effectively, it is useful to know the speed and extent of their dispersal and to understand the habitats they are using. Once habitat use in particular is accurately identified, these areas can be targeted for pest control. Knowing speed of spread enables control plans to be implemented that account for dispersal of the pest.

3. Marine animals are traditionally difficult to track because of the distance they travel and the vast expanse of the ocean. Radio-tracking, using satellites to detect the signal from the transmitter, has helped us to measure the distances covered by marine animals and the habitat they live in. In the case of the great white shark, data was provided on swimming speed and direction and showed that these animals do indeed undertake long migrations. This can help understanding of how these animals navigate. Radio-tracking can also provide information on nesting and breeding sites.

### Classification Keys (page 122)

1. The case (presence or absence and specific features of the case).

2. A *Oxyethira*
   B *Hudsonema*
   C *Olinga*
   D *Aoteapsyche*
   E *Hydrobiosis*
   F *Helicopsyche*
   G *Triplectides*

### Keying Out Plant Species (page 123)

1. (a) Silver maple, *Acer saccharinum*
   (b) Japanese maple, *Acer palmatum*
   (c) Red maple, *Acer rubrum*
   (d) Sugar maple, *Acer saccharum*
   (e) Black maple, *Acer nigrum*

2. Any one of: The size of the tree or shrub, the color of the bark and flowers, the shape of the winter buds and winged fruit.

3. Before a plant can be classified to species level a number of different features must be considered. One feature is often not sufficient to accurately distinguish between closely related species within the same genus.

4. Species must be correctly identified when carrying out population studies or it can lead to the wrong conclusions being made about a particular population (such population size, density, and distribution etc)

### KEY TERMS: Mix and Match (page 124)

Age Structure (R), Abundance (P), Belt transect (L), Density (E), Direct count (D), Distribution (G), Graph (Q), Indicator species (I), Indirect methods (B), Lincoln index (S), Line transect (F), Mark and recapture (H), Mean (C), Quadrat (N), Qualitative methods (K), Quantitative methods (J), Random sampling (M), Sample (O), Standard deviation (A)

### The Importance of Plants (page 127)

1. (a) **Food**: Plant tissues (leaves, fruits, nuts, seeds, tubers etc.) provide energy for heterotrophs. In some cases, fibrous or inedible portions can be used to make beverages through steeping (e.g. tea), or fermentation (e.g. alcoholic beverages).
   (b) **Fuel**: Plant tissues (e.g. wood) especially when dried, provide a source of fuel. Fossil fuels, particularly coal, formed from plant remains are also a rich source of fuel.
   (c) **Clothing**: Plants (e.g. cotton, flax, hemp) provide the fibers for a number of different fabrics including cotton, canvas, and linen.
   (d) **Building materials**: Plants provide wood (timber) and other fibers (e.g. fibrous grasses) for building fences, boats, and dwellings.
   (e) **Aesthetic value**: Plants create an aesthetically pleasing environment. People find plants pleasing for what they offer in terms of scenery, perfume, texture, and enhancement of personal surroundings. Examples include landscaped gardens with fragrant trees, public botanical gardens, and wilderness areas.
   (f) **Recreational drugs**: Plants are a source of mood altering drugs (e.g. marijuana, datura, kava), used for centuries by people for reasons usually associated

with their personal or spiritual lives. See also medicine below.
   (g) **Therapeutic drugs (medicines)**: Plants provide a rich, as yet not fully utilized, source of pharmaceuticals (e.g. aspirin from willow bark). Over 25% of modern medicines have been derived from plant extracts. Some plant poisons (e.g. digitalin) and some of the recreational drugs derived from plants (e.g. marijuana) may (with modification) also have medicinal uses in the modern world. More recently, the use of plant extracts as health supplements (e.g. St John's wort) or as alternatives to prescribed drugs (e.g. tea tree oil) has increased.

2. (a)-(c), in any order:
   – Loss of forests (particularly rainforest) represents a huge loss of biodiversity: the entire ecosystem is lost and recovery is slow and possibly not complete.
   – Loss of forests represents a destruction of a potential source of new foods and medicines.
   – With fewer forests, there are fewer global carbon sinks; carbon dioxide production rises and oxygen production falls. This contributes to global warming and all its associated biological consequences.

### Global Human Nutrition (page 128)
1. Subsistence and industrialized agriculture are on opposite ends of the spectrum with respect to their inputs of energy, labor, and capital. Subsistence agriculture is a low technology system, with minimal inputs of energy (these being largely in the form of physical labor). Enough food is grown to supply a family unit and only a small amount of land is cultivated as appropriate to this. In contrast, industrialized agriculture relies heavily on mechanization and the use of inorganic fertilizers, and it requires heavy use of (fossil fuel) energy to supply these. Industrialized agriculture requires high capital input (in machinery, fertilizers etc) but provides high yields per unit of land cultivated. Labor inputs are minimized because of the reliance on mechanization, but there is greater use of labor in supporting services (supply of machinery, fertilizer production, transport etc).

2. The modern industrialized agricultural techniques on which global food production is currently based are heavily reliant on fossil fuels, which are used to produce fertilizers and operate machinery. The decades following peak oil (when oil supplies peak and begins to decline), will see sharp increases in oil prices and consequently a rise in the costs of food production. This will cause food shortages through a number of interconnected mechanisms:
   – Increased costs of food production will increase food prices paid by consumers. This will disproportionately disadvantage poor nations where people will not be able to afford the increased costs.
   – More of the world's food crops will be converted to fuel production. There will be direct competition between fuel producers and food processors for supplies of wheat, corn, soybean, sugarcane, and other key crops.
   – Water deprived countries will divert water from irrigation and increase grain imports to feed their populations. This will reduce their internal (home-grown) supplies of food and make them even more susceptible to the fluctuations of the global market.
   – More countries with grain deficits.
   – More frequent regional famines are a likely further consequence of all of the above.

### The Green Revolution (page 129)
1. The first green revolution made use of increased use of machinery, and chemical fertilizers, pesticides, and water, as well as improvements in crop breeds in to increase the yield per unit land from key food crops. This greatly increased global food production but relied heavily on high energy inputs. The second green revolution has employed improved knowledge of plant breeding and genetics to produce high yielding strains with particular desirable qualities (such as disease resistance and drought tolerance) to increase yields further (particularly in tropical countries). Some of this second green revolution (especially more recently) has involved genetic engineering but some has been the result of improved selective breeding practices. This second green revolution is accelerating as a result of genome sequencing in crop varieties, although it is worth noting that continued gains in yield are frequently still reliant on high input agricultural techniques. The further step in the second green revolution is reduce reliance on chemical inputs.

2. (a) IR-8 rice was developed through selective cross breeding - choosing two strains of rice with desirable characteristics and breeding them together. Golden rice was developed by genetic engineering - artificially changing the genes of the rice plant by inserting genes from different organisms.
   (b) The advantages of the gene revolution include larger crop yields from the same amount of land and multiple harvests per season. This has helped solve many problems of global food shortages and slow the destruction of rainforest by reducing the need to clear more land for larger fields.
   (c) The disadvantages of the gene revolution include: many of the new crops require greater fertilizer and pesticide inputs than before. The larger and more frequent harvest require a greater input of fossil fuels and irrigation, affecting the biodiversity of the surrounding land.

### Cereal Crop Production (page 131)
1. (a) **First green revolution**: As a result of increased yield per unit of farmed land, achieved largely through high inputs of pesticides and inorganic fertilizers (together with improvements to crop varieties).
   (b) **Second green revolution**: Crop yields increased through the wide use of fast growing, high yielding varieties of crop plants. These varieties are specially bred for their suitability to specific regions, their high yield, and their pest and disease resistance.

2. Any of the following reasons: increased loss of arable land through soil degradation and erosion, increased resistance to pesticides used (relatively higher losses to pests despite increased pesticide use).

3. Wheat is the world's most important cereal crop. Common wheat contains a high protein content making it a valuable food resource. It is extensively grown in temperate area and is a major component of the economy.

4. (a) Sorghum is preferable to maize in regions prone to drought. The sorghum yield will decline when water supply is low but the crop will survive.
   (b) Most rice is grown for internal consumption (to feed the population of the nation in which it is grown).

5. (a) Any two of:
   – Stems of rice plants have large air spaces running the length of the plant. This allows air to penetrate the submerged roots.
   – Roots are very shallow, allowing access to oxygen dissolved into the upper layer of the waterlogged soil.
   – Cells can respire anaerobically when oxygen levels begin very low.
   (b) Any two of:
   – C4 pathway allows maize to fix carbon at low $CO_2$ levels.
   – Aerial roots help to prop up the tall stems.

### Chemical Pest Control (page 133)
1. Top consumers are most at risk because they eat a large number of prey items from lower trophic levels (therefore they consume larger volumes of the chemical).

2. Persistence gives a measure of how long a chemical will stay in the environment. Highly persistent chemicals are potentially more damaging because they last long enough to become incorporated into food chains. Chemicals with a very short half life have a good chance of degrading before they are taken up by organisms.

3. (a) 0.6 ng        (b) 2.0 ng

© 2007-2013 **BIOZONE** International
ISBN: 978-1-927173-60-2
Photocopying Prohibited

4. *C. quinquefasciatus* has a greater chance of developing resistance to permerthrin. The required dose for an LD 50 is higher than carbosulfan. The line on the graph also rises more slowly. This gives a greater possibility that a less than sufficient dose of the insecticide could be administered. Allowing more resistant mosquitoes to survive.

## Pesticide Resistance (page 134)
1. (a) and (b) any two in any order:
   - Insect populations tend to reproduce very quickly (generation times are short) and so the chances of a mutation conferring resistance (1) arising and (2) being passed on are greater.
   - Partial resistance can arise at several levels (e.g. behavioral, mechanical, biochemical) and through sexual reproduction in a given selective environment, offspring can require a sufficient number of mechanisms to develop full resistance.
   - Applications of insecticide do not always reach their intended target in the correct dosage and so do not kill 100% of the population, allowing surviving individuals to reproduce and pass on some resistances.

2. The insecticide application may not kill off the entire insect population. Those that survive will pass on any resistance they have to the next generation. Periodic insecticide applications act as a selection agent by allowing only the most resistant insects to survive and pass on their genes.

3. Resistance to synthetic insecticides has a number of implications. In order to maintain the kill rate of the insecticide, farmers often increase the amount, potency/, or toxicity of insecticide applied. This leaves more residue on the plant and so is potentially more dangerous to humans. Crops may have to be withheld longer before going to market and so cause loss of income to the farmer. Increasing resistance among insect vectors of disease is also a problem for human populations (e.g. resistance in mosquito vector for malaria), reducing the options for controlling the spread of disease through vulnerable human populations.

## Integrated Pest Management (page 135)
1. The aim of chemical pest control is to eradicate all insect pests, reducing the populations to zero. The aim of IPM is not to eradicate a pest but to reduce the crop damage it causes to an economically tolerable level (a level that still makes the crop production economically viable).

2. (a) Crop monitoring (before and during the control program) monitors the level of pest damage.
   (b) Cultivation controls combine crop management and physical means to reduce pest numbers.
   (c) If pest damage continues, biological controls, pheromone traps, and the release of sterile males are employed to control pest populations.
   (d) If pest damage remains at an unacceptable level, careful use of target-specific chemicals are used to control remaining pests.

3. (a) The principle of biological control is based on the reduction of pest species to a tolerable level using natural parasites, predators, or pathogens. It is rare, and not necessarily desirable, for the pest species to be completely eliminated (if the control agent is entirely reliant on the pest, then it too will be eliminated with elimination of the pest).
   (b) Precautions include the quarantine of potential biocontrol agents while their risk to the natural ecosystems is assessed. This assessment may be prolonged and involves investigating the control species biology so that predictions can be made as to how it will survive, carry out its function as a biocontrol agent, and interact with the organisms already present.

4. IPM allows the development of a more natural community on the farm, one which contains predators as well as pest and so can maintain the pests at certain levels. The reduced use of sprays reduces chemicals in the environment.

## Soil Degradation (page 137)
1. D

2. Salinization arises as a result of extensive clearing of the trees from both elevated and lowland areas, the introduction of new agricultural practices, the establishment of large scale irrigation, and the control and modification of river systems (e.g. no flooding).

3. (a) Nutrient enrichment: By organic farming which uses natural organic fertilizers such as manure and compost.
   (b) Pest control: Using biological control agents. Predators and parasites are used to keep pests at a low level. Management practices can also reduce pests. Planting species that repel insect pests. Sterilizing the soil using steam. Pheromone traps.

4. (a) DDT accumulates in higher concentrations in species at the top of the food chain. These animals appear to be affected. For example, predatory birds suffered from egg shell thinning which led to a massive drop in reproductive success.
   (b) Pest species are becoming resistant to the chemicals. The cost of research to discover and produce new pesticides is very high.

5. **Desertification** is the process of expansion or formation of drylands through loss of topsoil. The main causes of desertification include **deforestation**, **overgrazing**, and **intensive agricultural practices** such as high pesticide, herbicide, fungicide, and fertilizer use. Deforestation (the removal of natural land cover) leads to exposure of soil to wind and rain and increases erosion and soil instability. Tropical soils are particularly vulnerable because they are thin and nutrient poor. Overgrazing is a poor land management practice where high stocking densities on vulnerable land lead to soil compaction and loss of structure, as well as loss of topsoil. High chemical usage associated with intensive farming practices leads to chemical build up in soil and associated water courses and groundwater. Desertification may be arrested (or mitigated against) by stopping (or reducing the severity of) the activities that are causing the soil loss (e.g. removing livestock and farming trees instead). For example, in Kenya, beneficial tree planting in semi-arid regions provides wood for fuel and foliage fodder for cattle, and prevents soil erosion. Terracing and careful crop management can provide long term high productivity by reducing pesticide and fertilizer use, and maintaining soil health. In regions where salinization is the result of heavy irrigation regimes and flood control measures, returning the region to a natural cycle of flooding and drying can reverse the damage. This option is not favored because much of the land goes out of production during the process of recovery and thereafter. It also carries a high risk of inundation of riverside properties.

## Reducing Soil Erosion (page 139)
1. Windbreaks, contour plowing, intercropping, and terracing.

2. Planting parallel to the slope produces channels down which water can flow, taking soil with it. Contour plowing and terraces produce broad, flat areas which reduce water flow and prevent it carrying the soil away.

3. Maintaining vegetative cover reduces soil loss by reducing exposure of the soil to the elements. It also maintains the root stock of cover plants which stops the soil from be blown or washed away.

4. Intensive cropping systems tend to regularly expose the soil to the elements. This allows top soil to be lost. However, nutrients that are lost are generally replaced with fertilizers or crops that are plowed under. Alternative systems such as agroforestry and minimum tillage farming tend to keep the soil covered by grass of other plants for the majority of time. This reduces the loss of top soil and so also reduces the amount of fertilizer needed to replace nutrients lost through exposure of the elements. However, intensive cropping systems tend to produce larger more consistent yields than many alternative systems.

## The Impact of Farming (page 140)

1. (a) High input agriculture produces high crop and livestock yields. Mechanization reduces the need for many laborers and the use of pesticide use reduces crop damage. Disadvantages include the increased use of fertilizers, which has led to accelerated eutrophication (excessive nutrient enrichment) in many rivers and lakes. Pesticide use has potential negative health effects. The system has a high use of fossil fuels.
   (b) Sustainable farming practices reduce the need for pesticides and over fertilization. Less mechanization reduces the dependence on fossil fuels, reducing pollution and energy expenditure.

## Agricultural Practices (page 141)

1. (a) Pesticides are used to reduce pests and therefore increase yields.
   (b) Fertilizers increase the nutrients in the soil and so increase crop growth and yields.
   (c) Antibiotics are used to reduce the incidence of disease in animals living close together and thus reduce the risk of stock losses and epidemics.

2. The main energy uses come from the continual supply of added nutrients, pesticides, and herbicides (energy expensive to manufacture), and use of energy intensive machines (tractors, etc using fossil fuels) on the farm itself. Transport of materials and produce also consumes large amounts of energy in the form of fossil fuels.

3. Sustainable agriculture is based on the principle of long term production with minimal environmental impact. It centers around farm management practices that ensure good soil health.

4. Sustainable agricultural practices can prove to be more profitable in the long term because they release farmers from the economic treadmill of having to apply increasing amounts of water, fertilizers, and pesticides to the land in order to achieve the same yields. With improved soil health, high energy, expensive inputs can be reduced to a minimum with no reduction in yield, thereby saving money and increasing profits.

5. Near future agricultural practices are likely to be a combination of sustainable and intensive farming due to the growing cost (both economic and environmental) of using intensive farming methods. A complete change to sustainable methods is not currently possible because of the lower yields per unit of land. This would lead to food shortages and decreased yields. A combination of the two practices offers better economic and environmental outcomes in the future, while maintaining current yields.

## Forestry (page 143)

1. Strip cutting (in rapidly regenerating forest) or clear cutting in plantation forests (as these will be replanted on a rotational basis). Selective logging is not necessarily sustainable if large ancient trees are being removed.

2. **Clear cutting**
   Advantages: Easy to carry out. Requires little skill and planning. Cost effective.
   Disadvantages: Extremely destructive. Ecosystems can be wiped out. Recovery is limited and the system replacing the logged forest will be different in composition to the original.
   **Selective logging**
   Advantages: Reduced damage to environment. Reduces crowding of trees, encourages the growth of younger trees, maintains the age distribution of the original forest, can be sustainable.
   Disadvantages: Requires time and skill to select trees for harvest. Often expensive because it requires a helicopter to extract single trees.
   **Strip cutting**
   Advantages: Easy to carry out. Requires little skill and planning. Cost effective. Only a small area of forest is damaged.
   Disadvantages: Soil in clear cut land can erode while trees are regenerating. Cleared land could be invaded by aggressive plant species, causing changes to forest composition.

3. Old growth forests are important because they are biologically diverse and are often home to endemic and rare species.

4. (a) Ground fires burn underground (tree roots and peat) but can emerge at the surface. Surface fires burn along the ground on scrub and low trees. Crown fires burn higher up and engulf large trees. They burn extremely hot and cause large amounts of damage.
   (b) Controlled burns remove excess dry and flammable material, reducing the risk of wildfires (large out of control fires). Controlled fires also stimulate new growth and seed germination in some species.

## Managing Rangelands (page 145)

1. Grazing animals on the land helps to remove old material from plants. Removal of this material helps new growth by increasing the space and light available to the plant. Maintaining a grazing level that removes old growth without destroying new growth can increase total productivity.

2. Grazing opens up gaps in plant communities and can affect the types of plants that grow there. Plants that grow well with more light and space will tend to increase their range. The gaps also give opportunities for new species to invade. Intensive grazing may open up large gaps and remove many of the original species so that the species composition is changed.

3. Rangelands tend to cover arid and semi-arid lands. Plant communities living in these areas regenerate slowly because of the harsh environment. Removing plants by overgrazing can quickly turn these lands to desert, making future regeneration extremely difficult.

4. Primary production in a rangeland plant community in relatively low when ungrazed. Increasing the number of stock also increases the productivity of most of the plant types. This effect is only limited, however. Beyond a certain number of stock the plant material being removed by grazing becomes greater than can be replaced. At this point, productivity begins to decrease as the time between cropping becomes less, leaving no time for plants to recover. Overgrazing reduces new growing material and eventually kills the plant.

5. Rotating livestock through several areas of rangeland gives plants time to regrow damaged parts and produce new growth. This in turn helps to maintain the biodiversity by ensuring the rangeland is not overgrazed. Rotation of stock also helps to prevent build up of parasites in each area by breaking the parasite life cycle due to an absence of suitable hosts.

6. (a) Riparian zones provide habitats for many species due to their diverse character. They also help to reduce flooding by absorbing water and releasing it gradually, and reduce soil erosion.
   (b) Methods include: Watering holes for stock need to situated away from riparian zones to prevent their degradation by grazing, trampling and eutrophication. Stock levels should be reduced to limited grazing. Fencing could be used, although this can limit movements of other animals within the rangeland.

7. Over stocking can lead to overgrazing and the removal of a plants meristem, killing the plant. If this happens over too wide an area of an ecosystem for too long there will be no plants to replace those that were removed and no chance for new plants to grow. This leads to a complete change in the ecosystem and may lead to complete collapse.

## Reserve Lands (page 147)

1. National parks and reserves provide areas for *in-situ* conservation. Large populations and communities of animals can be managed and conserved in these areas in their

natural state while still providing access for recreation. At the same time, these reserves protect the environment that many outdoor activities require such as tramping, fishing, and kayaking. Another important feature of national parks is to provide an "advertisement" of a nation's best natural environments in order to attract tourists. National Parks with unique natural features, such as Yellowstone in the USA, and Fiordland in New Zealand, serve as focal points for tourists, bringing millions of dollars a year to the local economies. When these tourists return home, they then serve as advertisers for the parks they have been to. This in turn increases awareness of the importance of protected areas both at home and overseas.

2. Student's own discussion based on their chosen example. Protected areas usually have high conservation value because of the species found there (native, endemics etc), the esthetic values of the area, or the representative ecosystem type (for example, a remnant area of pristine habitat).

3. Private organizations are important in conservation for many reasons. Unlike government organizations, they are not funded directly by taxes and so do not compete with other government projects for funding. Funding can be gained from many sources, even donations from overseas, potentially giving access to far greater monetary resources and therefore to equipment for restoring and conserving habitats. Private organizations are often made up of volunteers for the most part (larger operations may pay a board of directors for their work) and can direct monies into operations other than wages. Private organizations also provide a supplement to government projects. Small operations may work on simple projects such as planting and restoring habitat while government departments manage more technical projects such as breeding and pest control.

## City Planning (page 149)

1. None of the city plans described perfectly matches a real city due to the fact that many cities have undergone different types of development throughout their history. As the city grows beyond its original limits new planning and positioning of areas may occur. The geography of a city also plays a part, with certain geological features such as rivers limiting the spread and size of some areas.

2. Disadvantages of urban sprawl include: environmental damage and overuse of infrastructure such as sewerage and roads. It can create problems to do with the extension of residential areas into industrial or rural zones such as noise pollution and reduction of farmland.

3. (a) Aspects that enhance social and economic aspects of a city include: quick, clean, and cost effective public transport, green spaces for public recreation, public entertainment and recreation areas, housing areas that provide quiet, safe and cost effective living, transport routes that reduce congestion.
   (b) Problems with converting existing cities into sustainable ones usually come down to money and infrastructure. Integrating new equipment into old designs can be extremely problematic. Electricity grids and power-lines often need to be redesigned and redirected. Building light rail and designing public transport routes to fit or match existing structure can be extremely expensive (and in some cases impossible).

4. (a) Transport will need to move people to within a short walk of their destination and be cost effective to use. This may be achieved by developing multiple rail lines in combination with buses. In order to reduce private vehicle usage, public transport needs to offer the public a wide range of schedules and destinations.
   (b) Cities can maintain an ecological balance by ensuring renewable energy is used to produce the majority of the power used. Greenways and green belts need to be established to give places of refuge for animals and building sites need to accommodate features in the environment, such as streams and stands of trees.

## Transportation (page 151)

1. Effective transport systems can reduce congestion by minimizing the need for private transport. This can be done by having fast and efficient public transport that delivers passengers to within a short walk of their destination. Public transport powered by renewable energy resources will help to reduce pollution and greenhouse emissions. Where private vehicles are required, road designs need to allow traffic to flow and change direction without the need to stop. This saves fuel, reduces congestion and reduces pollution caused by idling engines. Cycle lanes and greenways allow commuters to move about without the need for vehicles.

2. Private vehicles, buses, trains, and planes are all required to a certain degree in the transport of people and goods from place to place. All have aspects which are required to provide a comprehensive transport system. Private vehicles provide a flexible and personalized transportation system that is able to access places far from public transport routes. Buses can carry many people short distances between stops within a city, or over longer distances outside of a city. Trains provide both short distance transport within cities for large numbers of people and long distance haulage of freight and passengers between cities. Planes provide fast transport over any boundaries within and between countries, and can carry hundreds of people in one flight.

## Environmental Remediation (page 152)

1. Environmental remediation removes contaminants from a polluted environment in order to make it safe for human activities.

2. The benefits of remediation are numerous and include: recovery of resources from soils, such as metals; the restoration of soil so that it may be used again; the restoration of the environment to its natural form to encourage wildlife back into the area; improvement of health of the area, making it safe for human activities.

3. Students own answer based on their research. Methods that might be discussed include bioremediation, soil washing, thermal treatments, *in situ* and *ex-situ* methods.

## Mining and Minerals (page 153)

1. (a) Gold: circuitry, monetary policy, jewelry.
   (b) Iron: Building and construction (alloyed with carbon to form steel).
   (c) Gallium: Used as a semiconductor in microcircuitry and transistors.

2. Hubbert curves are theoretically symmetrical. Time to depletion can be estimated from the length of time from the start to the middle of the graph.
   (a) 45 years
   (b) 30 years

## Globalization (page 154)

1. (a) World bank: Provides financial and technical assistance to developing countries.
   (b) IMF: Improve global economic growth and stability. Provides financial help to member countries in difficulty.
   (c) UN: Maintains international peace and security by providing a forum where countries can talk about differences and solutions.
   (d) WTO: Helps the negotiation of trade deals between countries to improve import and export business.

2. Globalization has both positive an negative effects on the environment. Positive effects include a greater awareness of environmental issues, reduction of pollution by exchange of ideas and green technologies, reduce human impacts on the environment by exchanging ideas and research with a wider group of people. Negative effects include increased environmental degradation as countries develop and expand with new technologies (and so require more resources), confusion over different interpretations of shared data so that incorrect decisions or conclusions about the environment are made.

3. The Tragedy of the Commons can be applied to any resource exploited by more than one person or group. Examples on a local scale include the use of water from local wells or rivers, grazing animals on common fields of the use of timber from a forest. Examples on a global scale include fisheries in international waters, water use from large river and aquifers that span many countries, and even pollution of the air and water (the principle of individual gains causing wider scale losses still applies).

## Ecological Impacts of Fishing (page 155)

1. **Over-exploitation** refers to harvesting a commercial fish species so that the population falls below its optimal size. Overexploited populations show a progressive decline in growth rate and thus in population size.

2. **By-catch**: The part of the catch that is not the target species, or is discarded for economic, legal, or personal reasons.

3. The **maximum sustainable yield** (MSY) describes the largest amount of a naturally renewable resource (e.g. fish) that can be regularly harvested without causing a decline in the stock. A catch over this level will be more than the population can sustain. Reproductive individuals will not be able to compensate (by breeding) for the loss of biomass and the stock will collapse just as the anchovy fishery has collapsed. Once the anchovy numbers declined to low numbers, other fish species, notably sardines, increased in numbers (although their niches are not exactly the same). In the presence of high sardine populations, the anchovy populations cannot recover.

4. (a) At about 5 or 6 years, being the point where stock numbers are still moderately high, total fish biomass is high, and individuals are of an intermediate size (relative to maximum achievable size). If the few, older (larger) individuals are taken, the population quickly becomes skewed towards younger fish with lower reproductive capacity.
   (b) 0-6 years
   (c) Longevity, age at which reproduction begins, and mortality at different life stages.

5. Any three of:
   - Placing a size/age limit on take.
   - Enforcing and regularly reviewing maximum sustainable yield so that fish stocks can replenish themselves and the catch never exceeds what can be supported by the population.
   - Limiting the number of licences to fish issued.
   - Regulation of the fishing equipment used.
   - Limits on allowable by catch so that fishing vessels cannot keep fishing to remove only the large individuals while discarding smaller individuals of the same species.
   - For some species, supplementing the natural stocks with captive-bred fry.

6. Any two of for (a) and (b):
   (a) - Can be used to enhance natural fish stocks.
       - Can be used to take the fishing pressure off natural stocks.
       - Undesirable bycatch could be usefully used to produce fish meal.
   (b) - Producing fish meal to feed farmed fish uses more fish than is produced.
       - Fish farming can destroy natural fish habitat.
       - Fish farming can pollute as a result of large scale effluent flows.

## Fisheries Management (page 157)

1. Total landings and spawning stock biomass (a measure of the number of adults breeding) have been steadily declining since the early 1980s, and were indicating a decline (although not consistently) prior to that. As a consequence of declining spawning biomass, recruitment at age 1 also declined steadily during this period (with the exception of better years in 1984 and 1986). These data indicated unsustainable catches and decline of the stock below safe biological limits.

2. Summary responses only given:
   (a) North Sea
   (b) Risk of stock collapse is high. Stock is outside safe biological limits; spawning stock supported by only a few age groups and is less than half the level considered safe. TAC now half of the TAC in 2000.
   (c) Important features of biology: size (age) at harvest, breeding rate, age at first reproduction, growth rate, spawning behavior, effect of fishery on habitat.
   (d) Methods to assess sustainability include: surveys to estimate biomass (trawl and acoustic surveys, tagging), otolith examination to assess population age structure, assessment of stock recruitment (spawning assessment and survivorship).
   (e) Management options: size limits, deterring directed fishing, reducing by-catch of cod in other fisheries, restricted and closed seasons, reduced quota, closed areas (e.g. in breeding grounds), updating biological information on species spawning.

3. Without accurate estimates of age, population size and growth rate it is impossible to calculate an accurate MSY. Incorrect calculation of the MSY may lead to over fishing and the collapse of the fishery, or to under fishing with not enough fish being landed to create a viable economic resource.

4. (a) Over estimating the population size leads to overestimating its rate of replenishment and the size of the catch that can be landed.
   (b) Overestimating the growth rate leads to overestimating the size of the catch that can be landed due to the belief the population will quickly recover its loses.
   (c) This scenario could lead to the MSY being set too low due to the belief that the population ages and replaces its loses slowly. It may also lead to the belief that the population is close to collapse.

5. The statement is correct in that the population can only be harvested at the MSY if the population growth rate remains stable. If the rate drops but the MSY remains the same then each season a greater proportion of the population will be taken. For example a population of 100 has 40 taken each season (leaving 60). If a bad season sees the population recover to only 90 before the next harvest of 40 the population will be reduced to 50. A second bad season may see a further reduction. Harvesting at the MSY leaves the population vulnerable to changes in its population growth. Because of this, quotas are normally set at below the MSY.

## KEY TERMS: Mix and Match (page 159)

Aquaculture (R), By-catch (G), Chemical-pest control (V), Cultivation (T), Desertification (E), Globalization (C), Green revolution (K), Integrated pest management (P), Macronutrients (H), Maximum sustainable yield (J), Mineral (U), Pests (F), Rangeland (I), Remediation (Q), Soil conservation (S), Soil degradation (M), Stock indicator (N), Subsistence farming (O), Sustainable agriculture (B), Sustainable forestry (D), Urban development (L), Weeds (A)

## Using Energy Transformations (page 161)

1. Commercial electricity is normally produced by the rotation of an electromagnet inside numerous wire coils. The rotation of the electromagnet is accomplished by using a turbine turned by steam, water or wind. Electricity can also be produced for photovoltaic cells, although this is not usually used for commercial operations.

2. Energy is lost during each transformation, usually as heat. To turn the energy of falling water into electricity requires a number of energy transformations so that a large proportion of the energy in the water is lost due to friction and heat by the time the electricity reaches the home.

3. (a) Geothermal power: Heat in the Earth → kinetic energy in water → kinetic energy in turbine → electricity → heat lost in wire
   (b) Coal fired power station: Chemical energy in coal → heat → kinetic energy in water → kinetic energy in turbine → electricity → heat lost in wire

(c) Nuclear power station: Nuclear energy in atoms → heat → kinetic energy in reactor water → kinetic energy in turbine water → kinetic energy in turbine → electricity → heat lost in wire

Note: Each sequence may have other steps depending on where the sequence is started and where it is ended.

## Non Renewable Resources (page 162)

1. Coal, oil, and natural gas are all non renewable because the regeneration time of these resources takes millions of years. They will be used far faster than they are produced by nature. Once natural stores are used no more will be available to humans for millions of years. There is a fixed amount of uranium on Earth as it is formed during the super nova of heavy star. Large amounts reside inside the molten layers of the Earth but these are impossible to reach. The reserves on the Earth's surface are all that is accessible to humans and they cannot be replaced or replenished.

2. Issues associated with resource extraction:
   - Methods of extraction: these can cause large disruption of land by above ground mining, or land subsidence by below ground mining.
   - Resources for useful energy generation are not evenly distributed about the globe. Some areas are resource rich, others are resource poor.
   - Issues are associated with the safety and storage of nuclear fuel.
   - Energy resources in poorer nations as often controlled by wealthier countries or companies.

## Coal (page 163)

1. (a) Surface mining is used when the resource is close to the surface. The advantage of a surface mine is that large heavy machinery can be used to extract the resource quickly. Disadvantages include the disruption to the environment and the undesirable visual impact of the mine.
   (b) Subsurface mining is used when the resource is located far beneath the ground. Advantages include being able to reach the resources with causing large environmental damage and surface disruption. Disadvantages includes the difficulties of getting to, and extracting the resource. Gases may build up underground and specialized equipment is necessary to ventilate the mine. Underground fires may occur which can be extremely difficult to access and extinguish.

2. Environmental effects of mining include: leaching of heavy metals or toxic compounds into waterways, disruption of land (including erosion), noise and visual pollution, and air pollution by fine particles and exhaust from heavy machinery.

3. Anthracite has been compressed and heated for a longer period than peat. This has removed any moisture from the anthracite and concentrated the carbon. This results in anthracite burning more cleanly and producing a hotter flame than peat.

4. Coal provides a compact, easily obtainable, transportable, and storable energy source with a high net gain of energy. There are large supplies of coal, although these are non-renewable. However, the use of coal produces vast quantities of carbon dioxide, soot and sulfur dioxide (which contributes to acid rain). There is also a high level of land disturbance involved in its extraction.

5. In the United States (and most other developed countries) coal is used mainly in power stations. Little waste gases are produced because high grade coal if often used (which burns cleanly), and the power stations use sophisticated waste extraction equipment to extract waste gases. In the US, coal is not used directly by the majority of the populace in any great amount.
   In developing countries, coal is often used in open fire places in the home for cooking and to provide heat. This means people are more directly exposed to the toxic substances produced by burning coal. Power plants in developing countries often use low grades of coal (which produces more waste gases) and have less sophisticated extraction systems, producing more soot and toxic gases.

## Oil (page 165)

1. Natural gas is formed from hydrocarbon with four or fewer carbon atoms, e.g. $CH_4$, $C_2H_6$, $C_3H_8$, $C_4H_{10}$. Oil is formed from hydrocarbons with five carbon atoms or more, e.g. $C_5H_{12}$.

2. Oil is formed from the remains of algae and zooplankton which settled to the bottom of shallow sees and lakes about 300 million years ago. Over time these remains are covered with sediment and compressed. The organic substances in them meld to form oil. The process is extremely slow, taking hundreds of millions of years. So although the process is continual, the oil formed can not be replaced anywhere near the rate at which it is used. Thus it is classed as a non-renewable resource.

3. Advantages of using oil include: There is a large supply, oil supplies a high net energy gain when used as a fuel, crude oil can be refined and used to make many different products. Oil based fuels are relatively easy to transport. Disadvantages include the cost of extraction, the high amount of $CO_2$ produced from their combustion, and its limited supply.

4. Peak oil is important in forecasting the rise and decline of oil production. Peak oil represents the maximum oil production for any given oil well or reserve. According to the Hubbert curve, peak oil occurs at the halfway point of total production. When maximum production is reached, it can be estimated that half of the available resource has been used and so the time to exhaustion of that resource can be calculated.

5. Crude oil is placed in a distillation tower and heated. Short chain hydrocarbons boil off quickly and rise to the top of the tower where they are collected. Longer chains rise do not rise as far up the tower and are collected at different levels depending on their condensation point.

6. Butane is used in portable lighters because it is easily compressed to a liquid and can be contained in vessels with thin walls. It readily converts back to a gas at standard air pressure which can then be ignited easily. Thus a large fuel store can be compressed into a small space. Propane requires more compression and thus requires heavier, thicker walled vessels for its containment. As a result it is useful in gas stoves, heaters and barbecues that are not moved often. Propane returns to gas form more easily than butane and can thus be used at lower temperatures. Being gas, both hydrocarbons can be easily fed though pipes to burners that burn the fuel continuously. Longer chained hydrocarbons contain far more energy per unit than short chain hydrocarbons, thus petrol and diesel produce large quantities of energy from small volumes of fuel, useful in running the small motors of can, buses and trucks. Because these fuels are liquid at normal pressure they do not require any special vessel to contain them and thus can be easily transported and stored. When used in vehicles they work best when vaporized and thus require special injectors (or carburetters), but their high volatility (especially petrol) makes this task relatively simple. The lower volatility of diesel and longer chained hydrocarbons, makes them safer to handle and store, but requires them to be heated before being used. However they then produce huge amounts of energy per unit.

## Oil Extraction (page 167)

1. How the oil is extracted and exploited depends on where it is found and the type of oil or gas involved. Drilling rigs are used to access both land and marine based oil reservoirs, and the oil is pumped to the surface. Marine based oil reserves require floating or anchored oil rigs to be used. Once the well is drilled, the pressure in the reservoir forces the oil to the surface until the pressure in the reservoir is equal to the pressure above it. Natural gas or $CO_2$ can be injected into the reservoir to force more oil to the surface.
   **Extra note**: Oil bearing rocks are found by echolocation (bouncing sound waves off different layers of rock), measuring gravity fields or magnetic variations in rock layers,

and sampling rocks to identify oil bearing rock types.

2. The energy required to extract oil must be less than the energy gained from the oil that is extracted, otherwise there will be a net loss of energy.

3. (a) A well is drilled horizontally from the main well (up to 1500 m). The rock along this well is perforated (fractured). Fracking fluid is pumped into the well to keep the fractures open and this allows the oil and/or gas to flow back up the well.
   (b) Hydraulic fracturing increases the flow rate of oil and gas by fracturing rock that is impermeable, and increasing the pressure within the well by pumping fracking fluid into the well.

4. Conventional oil is oil that flows from an underground reservoir and requires little or no processing to make it into liquid form. Non-conventional oil requires extraction techniques other than those required for conventional oil. Non-conventional oil is found as a more solid form bound within sands and shales, and it requires heating and processing to liquefy it.

5. (a) 28.3 years
   (b) 47.2 years
   (c) These figures may be inaccurate as new oil resources are likely to be found, and renewable energy resources are likely to supplement and eventually overtake the use of oil. This will extend the lifetime of these oil resources.

6. Non-conventional oil requires a lot more energy to extract and refine than conventional oil. There is a lower net energy gain and more pollution associated with the extraction and refining of the non-conventional oil.

7. Offshore Arctic oil fields may soon be exploitable due to the shrinking of the Arctic ice-sheet, making exploration of the subsea floor possible. Extraction of oil from the Arctic could exacerbate the problem of ice sheet retreat due to the greater amount of fossil fuel that would be available to burn (producing more $CO_2$ and adding to anthropogenic global warming).

## Environmental Issues of Oil Extraction (page 170)

1. Both hydraulic fracturing and the mining of oil shales produce large volumes of liquid tailings that must be stored and treated in tailing ponds. Leakages from these ponds could cause contamination of ground water and water supplies.

2. Off shore oil wells are typically very deep. Any blowout or spill from the well itself would be very difficult to control at such depths, as demonstrated by the Deepwater Horizon and the Ixtoc I spills. The longer the spill takes to control, the more the oil could spread. Oil spilling directly from a well also has an large potential supply, as opposed to a fixed and smaller supply from an oil tanker.

3. Oil from *in situ* extraction produces more $CO_2$ per barrel because it requires more energy for its extraction and refining. The energy comes from burning fossil fuels and thus more $CO_2$ is produced.

4. Student centered discussion. Discussions might center around the gain in energy (a more energetic and easily transportable fuel is produced). Environmental issues might center around the amount of $CO_2$ produced and the effect on the landscape.

## Nuclear Power (page 171)

1. (a) Energy in a nuclear power plant comes from the fission (splitting) of uranium atoms. As the atom splits to produce two different atoms, some mass is lost as heat energy.
   (b) Heat energy produced during this process is used to heat water (or molten salts) in or surrounding the reactor. The heat from these is passed through a heat exchanger to produce steam to drive a turbine and generator. The nuclear reaction is control by inserting control rods that absorb neutrons in between the fuel rods. The more control rods inserted, the slower the reaction proceeds.

2. (a) Nuclear power generation has many advantages. It emits virtually no greenhouse gases and is capable of producing, from some calculations, almost limitless amounts of electricity. Reactors also require very little land use. The disadvantages of nuclear reactors include: the risk of a catastrophic accident similar to Chernobyl or the Fukushima Daiichi emergency. Although the safety systems on modern reactors reduce the risk of an accident, it only takes one accident (or natural disaster in the case of Fukushima Daiichi) to cause a colossal amount of damage. Other disadvantages include high start up costs and the possibility of refined nuclear fuel being used for weapons (both full nuclear warheads and dirty bombs).
   (b) To many people, nuclear power plants are linked to the production of weapons, massive environmental disasters, and genetic deformities. Knowledge that even a small spill of radioactive material can cause long term effects also affects peoples views. Much information is based on well meaning but often misinformed information from environmental groups. In reality, nuclear power production is extremely safe, but unlike other power stations, there is the potential for small problems to very quickly turn into enormous disasters. Many systems are also not designed to withstand massive external trauma (such as an earthquake) which may destroy backup systems (as in the Fukushima Daiichi emergency).

3. (a) 150 years   (c) 64,286 years
   (b) 300 years   (d) 7000 years

4. Nuclear fuel rods remain radioactive after use due to the fact that not all the fissionable material in them is used during their time in the reactor. While in the reactor, they also produce new radioactive material. Currently there is about 50 times as much energy left in a spent fuel rod as was used when it was in the reactor. However, they are no longer used because the fission reactions are no longer efficient enough to be useful in producing steam in the reactor.

5. To a certain degree the statement is true. Once installed, and during its working life, a nuclear reactor produces little or no greenhouse gases, produces constant and consistent high electrical power output for little cost, and if managed correctly and carefully is unlikely to produce any environmentally damaging waste. However, construction and decommissioning of the power plant is extremely expensive and this cost must be factored into the actual cost of electricity production. The construction and decommissioning also produces large amounts of pollution (both radioactive and non-radioactive). Spent fuel is also highly radioactive and needs to be carefully stored to prevent any effects to the environment. In the event of poor maintenance, safety, or an accident, nuclear reactors have the potential to produce large and long lasting environmental damage.

## Renewable Energy (page 173)

1. (a) Solar: High sunlight environment with few cloudy days.
   (b) Wave: Area with waves that are regularly spaced and of a regular height. Few storms or strong currents.
   (c) Wind: Area with consistent wind speed. Hill top or ridge.
   (d) Hydro: River valley with a narrow exit moving water from a high elevation to a lower elevation. High volume of water movement.
   (e) Geothermal: Steam-fields, volcanic activity close to surface.

2. Renewable energy is likely to become the predominant form of electricity generation in the future for a number of reasons.
   – It releases little in the way of greenhouse gases so will be more acceptable with the public.
   – Oil and coal prices are rising making them an expensive resource. Renewable technologies for the most part do not require fuels to power them.
   – As technology develops the cost of installing many renewable technologies will be reduced.
   – As the technology develops, generation efficiency of electricity will improve, increasing electricity production.

## Wind Power (page 174)

1. Wind conditions can change quickly and therefore affect the power output of the wind turbine. When wind power makes up only a small percentage of a country's electricity production, these changes in power output are generally indiscernible, as power from other generation stations make up the difference. However, when wind power makes up a larger proportion (above ~ 20%), these fluctuations make it difficult to match electricity output to demand.

2. Wind power provides emissions-free energy that can be easily located in many places and takes up little ground space once installed. However, because it is subject to wind conditions, wind power requires other energy systems to provide a base load. Some noise and visual pollution may also be produced. Wind farms need to be positioned carefully so as to not interfere with the flight path of flying animals.

3. (a) 11
   (b) $25.3 million
   (c) 53

## Hydroelectricity (page 175)

1. (a) Water builds up behind a dam and is channelled through tunnels (penstock) to the turbines. These drive generators which produce the electricity.
   (b) The larger the volume of water and higher it is stored, the greater potential energy it has and so that greater amount of electricity can be produced.

2. (a) Because they have only small dams, small scale hydro schemes do not produce large reservoirs that disrupt the flow of the river and therefore reduce interference with fish migrations, natural flooding events, or other natural river processes.
   (b) Pumped-storage effectively adds another power station to the grid during times of high demand by letting stored water flow through turbines in the same way as a normal hydroelectric dam.
   (c) Pumped-storage is an efficient use of electricity as it uses excess electricity during low demand times. Baseload power stations, which operate more efficiently at peak load, can be used to power the pumps to move water to the pumped storage reservoir. This allows energy to be effectively stored for later use during high demand times, and puts less demand on more variable power supplies.

3. The advantages of large scale hydroelectric dams are: They produce large amounts of renewable energy with only small amounts of greenhouse gas produced once rotting of vegetation in the reservoir has finished. The dams themselves provide flood control which prevents economic loss and loss of life. The reservoir can be used for recreation and for irrigation, allowing surrounding land to be farmed and produce valuable crops. Disadvantages include: High construction costs, river diversions required, interference with fish migrations, natural flood patterns are altered, and river valleys are drowned by rising water levels.

4. The statement is essentially correct if referring to the final product of the damming operation. (That is the completed dam running at capacity with no further flooding of the reservoir). However, while building the dam, hydroelectric power could be considered extremely environmentally unfriendly. The river requires damming and diversion, the upstream valley must be flooded, and people require resettlement. The $CO_2$ released during these operations is vast and the dam must run for many years before its economic and environmental costs are recovered.

## Solar Power (page 177)

1. (a) Advantages of solar power include: extremely low greenhouse emissions, high net energy gains, unlimited energy source in fine weather, and portable solar panel technology.
   (b) Disadvantages of solar power include: large amounts of land area required, high sunshine hours required, and low power output during winter, back up systems may be required.

2. Solar energy could potentially provide limitless power because the Sun is an external power source (i.e. its power is not generated by a resource produced and consumed on Earth) and its power output per day is many more times that humans use in a year. If this power could be efficiently harnessed, it would be able to provide all the energy we use now and into the future, even with growing energy demands.

3. Solar energy can be used to produce electricity at night by storing energy in a heat sink (or a battery). The heat sink may be stored molten salts that retain their heat for many hours after initial heating. These can then be used to heat water to steam at night and so drive turbines even when the Sun has set.

4. Passive solar heating uses no moving devices to heat a space i.e. there are no pumps used. Heat from the Sun heats the space directly and heat is stored in objects like concrete walls and floors. Active solar heating uses pumps to circulate solar heated fluids around the space. This allows water and radiators to be constantly heated.

5. A combination of active and passive solar heating, along with good insulation can help a house be energy independent. Photovoltaic cells could be used to provide direct electricity and charge batteries for electricity use at night. Active solar heating can heat water and provide some internal heating, while passive solar heating can provide the majority of warmth required during the day. A well insulated house will over-night retain the heat it acquired during the day.

## Geothermal Power (page 179)

1. Geothermal power is only a viable option in places where there is geothermal activity close to the surface and near urban centers. Only a few locations in the world meet these requirements.

2. Geothermal power production removes water from the ground. The pressure of this water as steam helps to support the surrounding land. Removal of the water without reinjection can cause both the subsidence of surrounding land and the depletion of the geothermal reservoir.

3. Steam is extracted at a near constant rate allowing geothermal plants to operate at near capacity for the majority of the time. This means they produce a constant and reliable electrical output.

## Ocean Power (page 180)

1. Equipment needs to:
   - withstand storms,
   - withstand strong water currents which may carry damaging sediment and debris,
   - survive the corrosive effects of seawater
   - produce constant electricity under variable conditions.

2. The energy stored in the oceans is enormous and renewable. It requires no input (or even management) from humans. Efficient and reliable energy harvesting designs could produce virtually endless energy supplies.

3. Technical problems surrounding ocean power designs and the lack of suitable sites mean ocean power is unlikely to produce significant amounts of energy in the near future. Ocean conditions can be unpredictable while storms and flooding can easily damage floating ocean power stations.

## Biofuels (page 181)

1. (a) **Biogas**: A mixture of methane (50-80%), carbon dioxide (15-45%), and water (5%) produced by the anaerobic decomposition (fermentation) of organic waste (e.g. sewage sludge or crop residues).
   (b) **Gasohol**: A blend of petrol and fuel alcohol (usually ethanol but sometimes methanol). The ethanol is produced by the fermentation of crop residues or other low cost sources of carbohydrate.

2. Corn ethanol is not a viable alternative fuel for several

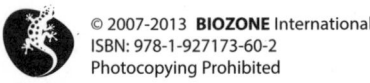

© 2007-2013 **BIOZONE** International
ISBN: 978-1-927173-60-2
Photocopying Prohibited

reasons. The corn required to produce the ethanol competes directly with the corn required as a food supply for humans and livestock, driving up food prices and reducing food supply. It also requires a large amount of energy to produce ethanol and the ethanol contains less energy than the equivalent amount of gasoline. Even using corn waste may have negative effects, because the waste normally would be used as fertilizer and this reduces the amount of synthetic fertilizers that need to be added to cropland.

## Current and Future Energy (page 182)

1. Global energy demand has grown at steady rate over the last thirty years.

2. Demand comes from an increase in the world population, the use of more energy-demanding technologies, and a growth in industry in developing nations.

3. Solar power:
   - The use or relatively rare semimetals will need to be overcome.
   - Efficiency of electricity production will need to be increased.
   
   Hydrogen fuel cells:
   - Production of hydrogen from water will need to be less expensive and energy intensive.
   - A lack of hydrogen fueling stations and infrastructure will be need addressed
   
   Nuclear fusion:
   - Ability to maintain reaction needs to be further developed.
   - Ability to control reaction needs to be further developed.
   - Currently requires more power to start, control and maintain than it produces.
   
   Space based solar collectors:
   - Technology required to build, transport and deploy is prohibitively expensive and currently impossible.
   - Many new types of technology would need to be developed for SBSCs to operate including launch and maintenance technologies.
   
   Nanomotors:
   - Require new molecular engineering techniques.
   - The ability to transfer the energy gained to a device would need to be developed.

## Energy Conservation (page 183)

1. Energy saving methods in transportation include:
   - improved manufacturing and design for better fuel efficiency in vehicles.
   - more use of public transport.
   - including a wider range of vehicles in the CAFE regulations.
   - use of hybrid vehicles.
   - development of alternative technologies e.g. fuel cells.
   
   Energy saving methods in buildings include:
   - Reducing energy leakages (through insulation, and double and triple glazing).
   - Environmental friendly design (orientation for passive heating and cooling).
   - Adequate ventilation to draw out moisture in winter and heat in summer.
   - Use of energy efficient appliances and lighting.
   - Installation of solar heating devices and heat exchangers.
   - Superinsulation and environmentally-friendly choice of building materials (earth, straw etc).

2. (a) 12
   (b) 1841.8 BTU per passenger kilometer

3. Building new power stations requires the expenditure of large amounts of money, the disruption of the environment, uses more resources and creates more pollution. None of which are currently viewed as costs worth incurring when a minor adjustment of people's electricity use habits and a slight increase in efficiency could save as much energy as a new power station produces.

4. (a) Incandescent: $41.67
   CFL: $20
   LED: $15
   (b) Incandescent: $600
   CFL: $140
   LED: $60
   (c) Cost to run existing incandescent bulbs:
   (41.67 + 600) x 20 = $12,833.4
   Cost to buy and run CFLs:
   (20 x 4) + ((20 + 140) x 20) = $3,280
   Cost to buy and run LEDs:
   (20 x 15) + (20 x 60) = $1,500
   
   Savings using LEDs in comparison to incandescent light bulbs: $11,333.4
   Savings of LEDs in comparison to CFLs: $1,780
   Note: the cost of replacing the existing incandescent light bulbs with new CFL or LED bulbs must be taken into account.

5. Students own discussions. Ideas may include:
   - The cost of the newer light bulbs is too high for many people.
   - The color and tone of the new bulbs is not agreeable to many people.
   - CFLs contain small amounts of mercury.

## KEY TERMS: Mix and Match (page 185)

Biofuel (K), CAFE (Q), Coal (F), Energy conservation (S), Fossil fuel (G), Geothermal power (T), Hydroelectric power (R), Hydrogen fuel cell (B), Industrial revolution (E), Natural gas (A), Non renewable energy (V), Nuclear fission (L), Nuclear fusion (U), Nuclear power (I), Oil (P), Photovoltaic cell (O), Renewable energy (D), Solar energy (C), Thermodynamics (J), Watt (W), Wave energy (H), Wind energy (N), Wind turbine (M)

## Types of Pollution (page 188)

1. (a) Pesticides/herbicides in waterways: Run-off and leaching from (intensively farmed) agricultural land.
   (b) Sewage: Discharge from urban centers.
   (c) Oxides of sulfur and nitrogen: Emissions from industry and power plants and vehicle exhausts.
   (d) Sedimentation and siltation of waterways: Construction activities and land clearance.

2. Urbanization increases the population load to a region and this is associated with increased use of resources and increased outputs of waste products, such as waste water, solid and hazardous wastes, and polluting air emissions associated with increased traffic volumes and concentration of industries.

## Water Pollution (page 189)

1. **Cultural eutrophication** refers to the increased enrichment of water bodies caused by human activity. It is caused by the runoff and leaching of excess nitrate and phosphate fertilizers from agricultural land.

2. (a) Domestic use: Shower/bath, washbasin, food preparation, toilet flushing, watering garden, cleaning car, drinking water, cleaning house (inside and out).
   (b) Industrial use: Cooling for plant machinery, cleaning, solvent for various processes.
   (c) Agricultural use: Irrigation, water for stock, cleaning (e.g. milking sheds and equipment), frost control (fine mist spray).

3. (a) Biological oxygen demand (BOD) is a measure of the polluting capacity of effluent. Specifically it is the level of dissolved oxygen available to microorganisms that decompose the organic material in the effluent.
   (b) Human activity such as discharging sewage rich in organic matter into the water, or agricultural fertilizers running off the land into waterways, contribute to excessive algal/aquatic weed growth. Some of this vegetation sinks and rots, thereby depleting oxygen in the water and increasing BOD.
   (c) In the short term, a high BOD causes death to the existing biota. In the long term it results in a change (reduction) in the community diversity in favor of

© 2007-2013 **BIOZONE** International
ISBN: 978-1-927173-60-2
Photocopying Prohibited

a small number of species tolerant of low oxygen concentrations.
   (d) If the sample were exposed to light then any living algae or plant organisms in the water sample will produce oxygen as a result of photosynthesis, thereby altering the dissolved oxygen levels.
4. (a) Liquid sewage is a cheap and valuable source of nutrients and can be utilized to promote the growth of crops.
   (b) Spraying effluent can lead to the contamination of land and food crops with fecal bacteria and viruses. Some microorganisms may cause diseases that present a health risk to livestock and consumers.
   (c) – Full treatment of the sewage to remove disease causing organisms and organic material. Water can then be used safely for a variety of purposes depending on the level of purity achieved.
       – Anaerobic decomposition of sewage sludge to produce biogas (methane).
5. Certain indicator species are indicative of particular aquatic conditions and can be used to detect changes in ecosystem status. Changes in macroinvertebrates over time, e.g. a shift from and diversity of clean water species to a smaller number of species that are more tolerant of poor water quality, can indicate a deterioration in the water conditions. Similarly, a return of clean water species following stream restoration can indicate when water quality has been restored to the required level.

---

### Sewage Treatment (page 191)
1. A = mechanical
   B = mechanical
   C = biological
   D = biological
   E = mechanical
   F = biological
   G = chemical

2. (a)-(g) Student's own summary and report. Summary responses will be very specific to the region. Problems of waste water management will center around cost (a major consideration as sewage treatment is paid for by landowners), availability of sites and public opposition to sitings, availability of suitable discharge points and public opposition to these, volume of waste water and capacity of current treatment facilities.

---

### Waste Management (page 192)
1. (a) Glass, paper, and plastic waste: Kerbside pick-up, commercial collection, and community recycling banks for recycling. Paper is recycled into cardboard/toilet paper. Glass is melted down and made into other glass products. Plastic waste can be recycled as packaging, lower grade plastic products, and fuel. All may also be dumped in landfill (this is undesirable).
   (b) Metals and their alloys: Community recycling initiatives for aluminum are strong in most countries. Cans can be recycled into new cans in 6-8 weeks. Steel is recycled into new steel products. The metals in vehicles are recycled (in part) into new vehicles.
   (c) Organic waste: Household or commercial composting. Compost can be sold profitably for garden and general use. Landfill disposal (an undesirable option).
   (d) Hazardous waste: Collected for incineration in high temperature furnaces. Some chemicals and radioactive materials must be sealed in containers and buried at secure landfill sites.

2. Hazardous wastes (e.g. radioactive wastes) are usually not part of an integrated waste management program.

---

### Reducing Waste (page 193)
1. Composting is useful in that it turns waste organic material, such as grass clippings, into a nutrient rich material that can be returned to the soil and so help to fertilize other plants. Beyond the initial set up (which may actually cost nothing) there is no cost and nutrients can continually be recycled from waste and returned to the soil to support the growth of other plants. Managed correctly, compost will never require too much space and virtually never run out.

2. Metals are acquired from ore mined from the ground and then processed. These are energy demanding activities. Reforming metal from waste metal removes the need for mining and processing. Often the metal needs little more than melting down in order to be used again.

3. (a) 0.2 MJ
   (b) 0.35 MJ
   (c) 0.25 MJ

---

### Plastics in the Environment (page 194)
1. Plastics are made using chemical molecules and bonds not often found in nature. As a result, bacteria have not evolved the metabolic pathways necessary to decompose the material and utilize the energy. Break down of plastics therefore relies on physical processes and takes much longer than decomposition of materials such as paper.

2. Where water forms eddies (areas of rotating water currents) floating objects (such as plastics)can be trapped and concentrated, producing large areas of debris.

3. Advantages of degradable plastics include: Breaks down in the environment, leaving little waste and reduces the need for recycling.
   Disadvantages of degradable plastics include: Break down of plastics in the environment may reduce the viability of convenient long term storage vessels (e.g. reusing drink bottles), and may increase demand required for the resources needed to manufacture new plastic products.

---

### Atmospheric Pollution (page 195)
1. Automobile exhaust (usually) or, if the city is a large industrial one, industrial emissions could be important contributors.

2. (a) Coal: Home heating, energy for power stations. Releases greenhouse gases and contributes to acid rain.
   (b) Diesel: Truck fuel.
       Green house gas emissions. NO is an ingredient in photochemical smog.
   (c) Natural gas: Home heating, car fuel, energy for power stations, industrial applications. Green house gas emissions.
   (d) Petrol: Car fuel. Green house gas emissions. Contributions to photochemical smog.

3. (a) **Biological indicators** are organisms, e.g. an aquatic insect such as a mayfly, or a symbiont, such as lichen, that are sensitive to various forms of pollution. Their presence or absence in an ecosystem is indicates whether the air quality if good or poor.
   (b) Some species of **lichen** are more sensitive to atmospheric pollution than others. The species found at a particular location can give an indication of the level of atmospheric pollution.

4. **Types of air pollutant**: Missing answers only are provided. In some cases, extra detail is provided.
   **CARBON MONOXIDE**
   **Sources**: Motor vehicle exhaust (forms when hydrocarbons are burnt with a limited supply of oxygen).
   **Environmental effects**: Poisonous to animals using hemoglobin to transport oxygen.
   **Human health effects**: Binds preferentially and strongly to hemoglobin, preventing it from transporting oxygen. Causes dizziness, headaches, impairs mental processes. Can cause death if exposed to high concentrations over short period of time. Cigarette smoke contains small amounts of carbon monoxide (CO) that chronically impairs the smoker.
   **HYDROGEN SULFIDES**
   **Environmental effects**: Will oxidize to become sulfur dioxide and contribute to acid rain.

**Human health effects:** Very poisonous gas, unpleasant odor.
**Prevention/control:** Use low sulfur or sulfur-free fuel.
SULFUR OXIDES
**Sources:** Coal-burning industries (including coal-burning power stations and metal smelters). Some coals have naturally low sulfur levels.
**Environmental effects:** Dissolve to form acids when in contact with moisture. Can cause severe leaf injury.
**Human health effects:** Irritate the nose and respiratory tract.
NITROGEN OXIDES
**Sources:** The exhaust from motor vehicle engines and high temperature furnaces.
**Human health effects:** Severely affect respiratory track, causing asthma and emphysema.
**Prevention/control:** Catalytic converters can be fitted to motor vehicle exhausts to reduce nitrogen and $CO_2$.
SMOKE
**Sources:** Carbon, soot and ash from motor vehicle exhausts, jet engines, fuel-burning power stations and industries, domestic fires, metal smelters.
**Environmental effects:** Reduces sunlight penetration and coats leaf surfaces (both reduce photosynthesis).
**Human health effects:** Creates a smoky haze. Aggravates respiratory problems.
**Prevention/control:** Remove smoke/dust particles from chimneys with electrostatic precipitators. Use alternative, non-burning energy sources.
LEAD
**Sources:** Emitted as lead oxide in the exhaust of cars using leaded petrol (in a few countries, tetraethyl lead is still added to petrol to increase octane rating and act as an antiknock agent. This practice has been discontinued in most westernized countries).
**Environmental effects:** Bioaccummulates in tissues and causes to lead poisoning. Interferes with functions within soil.
**Prevention/control:** Use unleaded petrol with a catalytic converter fitted to the exhaust.
OZONE
**Sources:** A secondary air pollutant formed by reaction between nitrogen oxides and volatile hydrocarbons (part of the process of forming photochemical smog).
**Environmental effects:** The most harmful component of photochemical smog, this tropospheric ozone reduces visibility. It also stresses plants and may contribute to forest decline.
**Human health effects:** It irritates and disturbs function of eyes, nose and lungs.
**Note:** Do not confuse this lower atmospheric ozone with that of the upper atmosphere where it forms a protective barrier to UV radiation. Tropospheric ozone does not replenish the ozone that has been depleted in the stratosphere because it is converted back to oxygen in a few days.
HYDROCARBONS
**Environmental effects:** React with other pollutant gases to form photochemical smog. Retard plant growth, causing abnormal bud and leaf development.
**Human health effects:** Carcinogenic.
**Prevention/control:** Ensure car engines are well tuned for complete combustion. Control vehicle exhaust emissions through the use of catalytic converters.

5. (a) **Sick building syndrome** occurs when a large office building is polluted by gases from equipment and microbes in the air-conditioning system.
   (b) Using low gas emission materials; treating the air conditioning system regularly for microbe contamination.
   (c) Up to 400 people sit in close proximity for up to 12 hours at a time. Contagious diseases can spread easily from one person to another, as well as through the cabin's air circulatory system.

## Cities and Climate (page 197)
1. (a) Cities contain a large amount of concrete and bare, reflective land that heats up the air above it during the day. Rural areas absorb more heat via vegetation and thus the air above them does not become as hot.
   (b) Cities tend to increase the rainfall downwind. Hot air above a city carries water vapor that condenses as rain when the air mass moves into rural areas and the air cools.

2. Inversion layers can stop pollutants from moving to higher altitudes and being carried away from built up areas. As a result pollutants, such as smoke, can build up at lower altitudes to dangerous levels.

## Acid Rain (page 198)
1. Acid deposition increases the acidity (lowers pH) of the water/soil that it falls into. This will affect the kinds of organisms that can live in the community as some will not be able to tolerate a drop in pH.

2. (a) Mussel (tolerates pH 6.0)
   (b) Water boatman (tolerates pH 3.5)
   (c) Determine which of these species is present in a lake (using a diversity index). The lower the diversity and number of species, the more adversely affected the lake. Also the specific species present will give an indication of the lake health.

3. Use sulfur-free fuels (low sulfur coal, natural gas) in industry, catalytic converters on car exhausts.

4. There are substantial costs involved in changing to a new fuel source or installing catalytic converters. Installation of costly equipment would need to be required by law (i.e. legislation is required) for people to carry it out.

## Noise Pollution (page 199)
1. Noise pollution can cause agitation, stress, insomnia and psychosis (usually due to insomnia). At intense levels noise can cause pain and hearing loss.

2. Noise pollution can be reduced by insulating machinery or using noise dampening equipment such as mufflers or anti-noise generators. Houses can be sound insulated. Industrial and residential areas can be separated to reduce noise pollution of homes.

3. Sources will vary but may include traffic, construction, trains, or airports.

4. (a) 100 x          (b) 10 billion x

## Toxicants in the Home (page 200)
1. (a) Student's own discussion. Responses will be specific to their own home environment. Discussion could include reference to sources of the pollutants in the seven groups described in the diagram.
   (b) Legislation by government is a primary measure in protecting people from contact with harmful chemicals. For example, PCBs and lead based products have been banned in many countries for many years. Consumer awareness is also important. Consumers should carefully read contents and ingredient labels when purchasing products.

2. People often relate pollution with the environment beyond the house, thinking of car exhausts, dumping of waste and factory emissions. People either forget or don't realize that pollutants can build up inside from products slowly decomposing and releasing chemicals.

## Health Effects of Pollution (page 201)
1. Pollutants that people are regularly exposed to may include:
   - CO: Headaches, respiratory damage
   - Fine particles (smoke, ash, dust etc): Respiratory illness.
   - $SO_2$, $NOx$, $O_3$: Cardiovascular disease
   - Fertilizers/pesticides: Skin irritations, gastrointestinal illness and vomiting.
   - Radiation (mostly solar. Radioactive waste is much less common): Skin cancer (melanoma)
   - Heavy metals: Headaches and nerve damage.

2. Social effects of pollutants include the disruption of

relationships as affected people may require care from parents, children, or partners. These people may be unable to work because of their care duties resulting in lost productivity to the economy. Health insurance or welfare payouts may increase. Treating diseases caused by pollutants may cost countries hundreds of millions of dollars a year (e.g. $150 million to treat lung diseases in the USA).

### The Economic Impact of Pollution (page 202)
1. Total removal of a pollutant usually carries a greater economic cost than the cost of incurred by the pollutant remaining. The value of the break-even point is to establish at what point the economic cost of clean-up exceeds the costs incurred by the cost (to society) of the pollution.

2. Example will be student's choice, and the costs associated will vary accordingly. Some costs are:
   (a) Direct costs:
      – cost of clean-up.
      – cost of salvage of equipment.
      – cost of species relocation or rehabilitation.
      – cost of public notifications.
   (b) Indirect costs:
      – costs of revenue losses.
      – cost of mitigation.
      – cost of loss of public confidence.
      – cost of loss of aesthetic values.

### *Exxon Valdez* Oil Spill (page 203)
1. (a) The *Exxon Valdez* oil spill could have been avoided if crew had had their mandatory rest periods and were not tired. The radar system should have been operational and functioning to help crew avoid the reef.
   (b) The spill caused serious oilling of the seashores along many hundreds of kilometers. It caused of the deaths of numerous marine species including birds, fish, sea otters, seals, and whales.
   (c) The cold temperatures in the region meant more volatile parts of the oil evaporated only slowly. Storms and high tides made keeping oil from the beaches difficult. Storms made using dispersants and controlled burning almost impossible. The remote location made it difficult to bring in equipment and supplies.

### Niger Delta Oil (page 204)
1. Environmental effects include:
   – Water contamination.
   – Fishing grounds polluted and fish populations dwindling.
   – Air quality affected by flaring and oil vapor.
   – Oil slicks cover water and ground surfaces.

2. Money from the oil companies paid to the government does not appear to be reaching the local communities who were affected. While government officials appear to be reaping the rewards, most of the communities living around the affected areas are still poor and lack basic facilities. Oil companies appear to be ignoring the problems caused by oil spills.

3. Conflict makes any job dangerous. Kidnapping of oil workers, and damage to pipelines have been carried out by some armed gangs operating within the Delta. Damage to pipelines is often slow to be repaired because oil companies are reluctant to send workers into dangerous areas. Conflict can lead to administration difficulties or corruption, making regulation difficult. This can lead to environmental issues when a company fails to follow guidelines (knowing there will be little government reaction). In addition, during times of conflict governments tend to focus more on restoring order and keeping the peace than on the environmental impact of any particular action. This means companies operating in an unstable area may be under less scrutiny than they would be in a stable area.

### Deepwater Drilling (page 205)
1. New technologies, including include larger thrusters for positioning, GPS, and larger, stronger drilling equipment, have allowed oil rigs to drill in deeper water.

2. Oil from wells in shallow water is beginning to diminish in supply. New technologies now allow oil companies to explore deeper water in order to find new oil reservoirs far more easily than was once possible.

3. Issues include:
   – Ocean currents are stronger, requiring larger thrusters to maintain position.
   – Equipment is heavier and there is more of it, requiring larger, stronger machinery to manage it.
   – Accidents at depth are more difficult to contain or fix due to high pressures and low temperatures.
   – Distance from shore is usually greater than in shallow water, making supply of vessels and retrieval of stored oil difficult and more costly.

### *Deepwater Horizon* Oil Spill (page 206)
1. – The cement and metal casing lining the well bore failed to keep gases in the oil from entering the riser.
   – The gases reached the surface and ignited causing a fire that eventually sunk the oil rig.
   – The riser ruptured causing a blowout at the wellhead.
   – The blowout preventer (BOP) failed to operate correctly and seal the well.
   – The automatic deadman's switch also failed to activate the BOP.

2. Estimates of flow rate varied due to:
   – The methods used to estimate the rate varied, and gave different estimates.
   – The operations surrounding the containment of the oil spill affected the amount of oil spilling into the Gulf.
   – Flow from the well may also have been affected by changes in pressure in the well itself.

3. The purpose of the blowout preventer (BOP) is to seal the wellhead in the event of an uncontrolled release of oil from the wellhead.

4. (a) 6500 to 19,500 times more oil was leaked during the *Deepwater Horizon* spill than the *Exxon Valdez* spill.
   (b) Oil reservoirs contain many times more oil than can be carried in an oil tanker. Therefore, oil leaks from drilling platforms have the potential to be much larger than from an oil tanker spill. A blowout at the wellhead has the potential to spill enormous volumes of oil while oil tanker spills, although large, are at least finite in size.

5. Humans cannot withstand the very high pressures and cold temperatures experienced deep underwater, so remote operated vehicles (ROVs) are needed to work at the wellhead. **Note:** Currently the deepest dive record for a human in a dive suit is 600 m, nearly 1000 m short of the Mocondo wellhead.

6. The containment dome failed because it became clogged with crystals of methane hydrate. Oil was also leaking from other parts of the broken riser pipe.

7. (a) The top kill method involved pumping heavy mud through a by-pass on the BOP into the well bore.
   (b) This method was an attempt to force the oil back down the well before sealing it. The attempt failed because the pressure required to do this could have caused further rupture of the well, which would have worsened the situation.

8. (a) The bottom kill method involved drilling a relief well into the blown out well and using this to pump mud and cement into the well bore, sealing it from below.
   (b) The blown out well was located by lowering a transmitter and sensor into the drill shaft drilled by the relief rig. The transmitter generated an electric field causing the casing of the well to produce a magnetic field. This could be detected by the sensor. The magnetic field increased in strength the closer the relief came to the blown out well.

9. Several attempts were used to give multiple chances of sealing the well. Less complex attempts were tried first (e.g. lowering the containment dome) in order to save time and money, but also to make sure that if the attempt failed it did not affect the ability to try other methods. For example; had

the riser pipe been cut first and the LMRP cap subsequently failed, the increased flow from the well would have made the oil spill even larger.

### Oil Spills and Wildlife (page 209)

1. Oil is a viscous fluid comprised of numerous types of hydrocarbons. Its viscous nature causes it to easily adhere to surfaces and makes it difficult to remove. The low density of oil allows it to float on water and therefore be carried along with water currents, and be moved about by wind blowing over the water's surface. Oil can also spread out to form thin slicks that cover large surface areas. Because it floats on water, as the water moves up a beach with the tide the oil is carried along with it. The oil sticks to the sand, rocks, or grasses along the beach and as the water retreats with the out-going tide, oil is left behind. Each tide can bring in more oil, increasing the problem.

2. Immediate effects of oil on wildlife include:
   - Fur and feathers become clogged with oil, reducing flotation and insulation.
   - Vapors cause breathing difficulties.
   - Ingestion causes intestinal illness.
   - Oil makes food sources inedible causing starvation.

3. (a) The effects of an oil spill depend upon the type of oil and the environment into which it is spilled. Because each environment is different and the conditions experienced in that environment change (through the weather) how a particular oil may act once introduced can not be fully predicted. The long term effects of an oil spill may be even harder to predict for the same reasons.
   (b) The magnitude of the ecological effect of an oil spill is not necessarily linked to the size of the spill because the type of oil and the environment play major roles in its degradation. Light oil and warm temperatures combine to cause the evaporation of much of the oil's volume, where as heavy oil and cold temperatures will slow this process down. The presence and actions of microbes in the water also has a big effect on the longevity of the oil. The density and type of wildlife within a spill area also plays a factor. Oil spilled in estuaries and enclosed bays with high concentrations of wildlife will have a greater effect than oil spilled in the open sea.

4. Oil spills are ecologically disastrous events because they affect a wide variety of wildlife. For the same reason they are economically disastrous. Fish and shellfish stocks are significantly depleted causing jobs related to fishing (fishermen, processing, and fish markets) to be lost. Tourism may also be affected. Other business are affected because the affected workers no longer have the funds to spend in the community. In communities where a large portion population works in places affected by the oil there can be significant and long term economic and related social effects.

### Cleaning Up Oil Spills (page 211)

1. Floating booms can be used for: containment, protecting coastlines; collecting oil for burning; containing oil while burning.

2. Oil is burnt in order to reduce its volume. The effects of this produces large volumes of dense smoke and carbon dioxide.

3. Dispersants stop oil from forming continuous slicks by breaking the oil into minute droplets that then dissolve in the water. This increases the surface area of the oil and speeds up degradation. Issues associated with dispersant use include long term effects of using the large quantities at such depths in the Gulf. These have yet to be fully determined.

4. Steam was not used in the Gulf because it has the potential to do more damage than good. The high temperature of the steam kills any organisms that are sprayed with it.

5. (a) Oil is broken down by microbes into other compounds.
   (b) Oil forms a mousse, a frothy glutinous foam that makes dispersal difficult.
   (c) Evaporation removes the smaller, lighter molecules from the oil, reducing its volume.
   (d) Oxidation causes the oil to form soluble compounds. Partially oxidized oil forms tarballs which may be deposited on the seabed and beaches.
   (e) Oil becomes heavier and forms deposits on the seabed or disperses in the water column.

### The Effects of Nuclear Accidents (page 213)

1. (a) The events that led to three of the reactors at Fukushima melting down were triggered by the 9.0 magnitude earthquake. The earthquake caused the reactors to shutdown and diesel generators to start up to run cooling pumps. The tsunami that followed destroyed the generators resulting in a shutdown of all cooling systems. This resulted in a loss of coolant in the reactor which caused the reactor cores to melt.
   (b) The explosions in the reactor buildings were caused by a build up of hydrogen gas which resulted from the exposure of the fuel rods to air.
   (c) The events at Fukushima can be called a man-made disaster due to failures in following basic safety procedures, a lack of preparation for such a disaster (even though Japan regularly experiences large earthquakes), the failure of regulators to properly monitor safety around nuclear reactors, and a failure of the operator to act on a warning that the backup generators were vulnerable to flooding.

2. (a) Although there are a number of spikes in radiation, the general trend is a rapid drop in radiation leaks from the plant.
   (b) 1 Seivert approximately equates to a 5.5% increase in risk of cancer. The highest radiation levels at the plant gate reached just under 4000 µSv (0.004 Sv). The level of risk in developing cancer in therefore still very low (although much increased from natural background radiation).

3. The Chernobyl accident release more radiation than Fukushima because the explosion at Chernobyl breached the containment vessel exposing the reactor core and the burning radioactive graphite lining the core. None of the containment vessels at Fukushima were breached.

4. (a) Radiation from the Fukushima plant has been detected in food (beef, rice, fish) and in waterways. There is evidence to suggest that populations of butterflies around Fukushima are experiencing higher than normal levels of mutation.
   (b) The radioactivity emitted by the explosion caused the death of people and wildlife from radiation poisoning. Mutation rates have increased in animals in the main fallout zone. Thyroid disease is higher than normal in both humans and other animals.

5. (a) Pike occupy a higher trophic level than non-predatory bream. Pike consume other fish species and concentrate radioactive particles in their tissues through the many prey items they eat (bioaccumulation).
   (b) The higher than normal levels of radioactivity in tuna caught off California are a result of radioactive material leaking from the damaged Fukushima power plant in Japan. Tuna is a top predator and will tend to have higher concentrations of radioactive particles in their tissues than non-predatory fish.

### Bhopal Disaster (page 216)

1. Approximately 38km². (Estimates of area affected vary, with some estimates covering up to 78 km²).

2. The plant was originally designed to produce a pesticide that was nontoxic and did not persist in the environment. This would have helped control pests and increase crop yields. The operation of the plant also created jobs for local workers.

3. Reasons for the disaster include (any two of):
   - The storage of hazardous intermediate chemicals. This danger could have been removed by ensuring the intermediate chemicals were immediately used.
   - Poor plant maintenance. This could have been avoided

if equipment was serviced and replaced regularly.
- Safety systems were insufficient. The plant should have been designed with greater safety in mind. Tighter regulations in the construction and design of the plant and regular safety checks should have been performed. Managers should have made sure safety equipment was service regularly and adequate for the job.
- The plant was built in a populated area. The plant should have been built in a less populated area. Housing built nearby after the plant was built should not have been allowed by local government.

### The Role of Legislation (page 217)
1. An Environmental Impact Statement allows for public comment and scientific evaluation of the possible impacts of a particular development (e.g. building a new hydro power plant). The EIA can be used to determine the possible positive and negative affects of a development.

2. (a) The purpose of Article 1(1) is to plainly and clearly state that smoke may not be emitted, and that emission of the smoke is an offense.
   (b) 1(4) provides the only exemptions from prosecution for the production of smoke, thus giving a reasonable margin for furnace operation.
   (c) Dark smoke is defined so that there is a measurable limit for the production of smoke.

### KEY TERMS: Mix and Match (page 218)
Acid rain (D), Atmospheric pollution (E), Biological oxygen demand (L), Direct costs (J), Eutrophication (C), Indicator organisms (I), Indirect costs (H), Integrated wastes management (B), Noise pollution (M), Non-point source pollution (A), Organic effluent (K), Oil spill (N), Photochemical smog (O), Point source pollution (G), Pollution (F), Recycling (P), Sewage treatment (R), Toxicant (Q)

### Models of Climate Change (page 220)
1. Climate change predictions involve many complicated parameters and will never be fully accurate because they are based on scenarios in which parameters can only ever be estimated. Some parameters are simplified or ignored in order to make the models easier to work with. It is also difficult to predict future human actions, and the effect of these actions on climate change. Therefore, scientists can only suggest a likely course of events; they can not provide a definite outcome.

### Global Warming (page 221)
1. (a) Carbon dioxide:    37.1% increase
   (b) Methane:    156.6% increase
   (c) Nitrous oxide:    18.7% increase

2. More frequent and prolonged seasonal flooding as well as permanent inundation of land. Increased coastal and inland erosion. Loss of small, low-lying atolls.

3. Increased levels of carbon dioxide, methane, and nitrous oxides act as additional blankets around the Earth, allowing the sun's energy to reach the Earth's surface, but preventing the heat escaping. This means that the Earth slowly heats up. **Note**: The atmospheric concentrations of these gases have increased dramatically above pre-industrial levels since 1750. These levels are considerably higher than at any time during the last 650,000 years (the period for which reliable data has been extracted from ice cores) and are correlated with a rise in global temperature and documented sea level rises. While correlation does not mean cause and effect, the majority of climate scientists accept the theory that the increase in anthropogenic greenhouse gas emissions is causing the rise in the Earth's temperature.

### Biodiversity and Global Warming (page 223)
1. (a) Global warming will result in an increased frequency of weather extremes (e.g. floods and droughts) and a loss of land as coastal areas are inundated. Erosion rates may also increase as a result. Glacial retreats will reduce water supplies and snow lines will increase in altitude. Climate changes may shift the governing physical environment in certain regions (and consequently cause a shift in predominant vegetation). Ocean pH will also fall as a result of $CO_2$ absorption (again, with consequent changes in biotic communities).
   (b) In general, crop growing ranges may shrink, expand, or shift. Crop plants may be affected more by higher night temperatures than by higher daytime temperatures. High night temperatures affect the ability of some crop plants such as rice to set seed and fruit. This will cause a reduction in the harvest, and a decrease in the amount of seed available for subsequent plantings.
   (c) Farmers may adjust by planting different crops in some areas, e.g. crops that are able to grow and set seed in the higher temperatures. New strains of crop plants may be able to be developed for the higher temperatures.
   (d) Migratory birds in the northern hemisphere are now not travelling as far south during the winter months (as higher latitudes become more hospitable) and they are making their migrations north up to two weeks earlier than usual.
   (e) Migratory birds may arrive at feeding grounds before the main food supply is ready. Plants with daylength-dependent flowering may not yet be flowering and, as a consequence, insects (and seeds and fruits) may not be in the abundance required to feed the migrants. In addition, the distribution of food resources may remain the same, but the birds are not migrating as far south and may be disconnected from their winter food supplies.
   (f) As air temperatures rise, so too does the snow line in alpine areas. Animals living on or above the snow line will be forced into smaller areas. If they are unable to move to higher latitudes where the snow line is lower, it is inevitable that they will become extinct in their native ranges as they run out of food and space.

2. Evidence suggests that insect populations will be affected by global warming. Butterfly populations in many areas have been recorded as shifting to higher latitudes and altitudes. Fossils of insect browse damage also suggest insect populations will increase in size as temperatures rise.

### Ice Sheet Melting (page 225)
1. Both low sea-ice albedo and area cause more heat to be absorbed by the land and sea. This heat causes the ocean's water to warm during the summer and therefore take longer to cool during the autumn (fall). This leads to winter sea-ice taking longer to form and being thinner than usual in the winter. Thin sea-ice has a lower albedo than thick sea-ice and melts faster in the spring leading to even less sea-ice the following winter. Thus thin ice and small area cover causes even thinner and less cover the following year in a potentially perpetual cycle until the sea-ice is lost.

2. Polar animals that live out on the sea-ice will be directly affected as the area covered by sea-ice reduces. Polar bears hunt out on the ice will find it harder to find food and will have to swim longer distances to find firm ice. The reduction of sea-ice cover may also have effects on species that live below the ice as more light and heat will penetrate the waters and to deeper depths.

### Global Warming and Agriculture (page 226)
1. Some crops (e.g. wheat, rice and soybeans) may benefit the higher temperatures and $CO_2$ levels. They may have longer growing seasons, or their growing region may expand. Other crops may be harmed by global warming. These include crops that are already being grown near their climate thresholds.

2. Climate change may allow pest species to expand their habitable range so that a wider variety of crops, or crops in previously unaffected areas, may be damaged by the pest.

## Temperature and the Distribution of Species (page 227)

1. Species distribution of *Rana* is closely related to water temperature (mating and embryonic development for each species occurs within a certain temperature range). Increasing global temperatures may result in a northwards shift of some species as their preferred water temperature shifts. Those frogs that are furthest north (*R. sylvatica*) may end up with a reduced range, while those further south (*R. clamintans*) may increase their range, depending on available habitat.

2. Climate change may cause changes in temperature enough that species may be able to (or have to) change where they live. Populations of animals in cooler climates may be forced north in the Northern Hemisphere, while species from warmer climates may expanded their ranges further north and south from the equator.

## Ocean Acidification (page 228)
1. D
2. C
3. B
4. A
5. pH and acidity are inversely correlated. As the pH decreases, the acidity increases.

## Carbon Trading (page 229)

1. Farmers and forestry can create carbon credits by growing plant material that takes in $CO_2$ from the air. If they crop these plants they must make sure the $CO_2$ output from their cropping is less than the $CO_2$ taken in by the plants and returned to the soil when ploughed in or cut down.

2. Carbon trading works by giving every tonne of $CO_2$ produced or saved in industry a monetary value, called a credit. Credits can be bought or sold on an open market with supply and demand determining the price of the credit. Companies producing more $CO_2$ than allowed must buy carbon credits to top up their $CO_2$ limit. Companies producing less than their limit can sell their excess credits for profit.

3. The advantages of carbon trading include allowing governments to cap the total $CO_2$ amount while allowing companies the flexibility required to compete in the market place. It also gives incentives for companies to be less polluting as they can then sell excess credits for profit. Polluting companies are forced to pay penalties to the market place by buying more credits which may harm their profitability. However the disadvantages of the scheme include very large companies selling high demand consumer products, such as oil, passing on the costs of their pollution to the public. Credits produced from farming or forestry are difficult to regulate and may in some cases not be equal to a full tonne of $CO_2$, causing more $CO_2$ to be released without companies realizing it.

## Carbon Capture and Storage (page 230)

1. Similarities: Both post combustion and oxyfuel combustion systems burn fuel then capture the $CO_2$ produced. Electricity is produced by heating water to steam that drives a turbine. Both use fuel in a pulverized form.
Differences: Precombustion captures $CO_2$ from the fuel before it is burned. (Fuel is gasified before use). $H_2$ is combusted to produce steam directly that drives the turbine. Excess heat heats water to drive a second turbine.

2. Captured carbon dioxide can be stored in deep geological formations or reacted with minerals to form carbonates, which can then be stored or used in manufacturing.

3. 
    - Fossil fuel fired power stations produce enormous volumes of $CO_2$. The $CO_2$ capture processes need to be huge to accommodate this. (Current prototypes can capture barely 3 days worth of $CO_2$ production from a power station per year).
    - The capture system requires power, which will be drained from the power station it is attached to.
    - Storage of $CO_2$ may be difficult. Deep geological formations may prove unstable in future. Injecting $CO_2$ into the deep ocean will raise the pH.
    - Reacting $CO_2$ with chemicals to form solids raises the question of what to do those solids.

## Stratospheric Ozone Depletion (page 231)

1. UV radiation is a powerful carcinogen causing an increase in mutation rates and generally interfering with genetic processes. Notable are increased rates of all types of skin cancers.

2. UV light causes the release of free chlorine from CFCs and this chlorine breaks down and destroys the ozone. The ozone layer absorbs most of the incoming UV and prevents it from reaching Earth. With fewer ozone molecules to absorb the UV radiation, its penetration through the stratosphere is much greater and more reaches the Earth's surface.

3. (a) Greatest geographical extent: September to early October (Southern hemisphere early spring).
   (b) Most depleted: Mid-October (1992).
   (c) Trend of ozone depletion: A steady decline over the last two decades with the exception of 1989 when there was a brief increase to 1983 levels.
   (d) In September 1997 the concentration of ozone increased between altitudes 10-20 km. By October 1997, the level of ozone had declined markedly (to approx. 0 mPa pp at 15-20 km) between these altitudes (the ozone was severely depleted at these altitudes in October but not September).

4. Development and implementation of new technologies required to reduce ozone depleting chemicals is costly. Furthermore, the technology is controlled by the affluent Western economies. Poorer, developing countries will find it difficult to spend large amounts of money on converting to alternative technologies. Recent studies suggest that some of the proposed replacement chemicals may also cause ozone damage.

## Loss of Biodiversity (page 233)

1.  1   **Tropical Andes**
        The richest and most diverse hotspot where it is home to 20,000 endemic plants and at least 1500 endemic non-fish vertebrates.

    2   **Sundaland**
        Some of the largest islands in the world are found here in Southeast Asia. The second-richest hotspot in endemic plants, and well known for its mammalian fauna, which includes the orangutan.

    3   **Mediterranean basin**
        The site of many ancient and modern civilizations, it is the archetype and largest of the five Mediterranean-climate hotspots (also see nos. 9, 12, 19 and 22). One of the hotspots most heavily affected by human activity, it has 13,000 endemic plants, and is home to a number of interesting vertebrates such as the Spanish ibex.

    4   **Madagascar and Indian Ocean islands**
        Madagascar is a top conservation priority as this 'mini-continent' has undergone extensive deforestation. This hotspot is famous for reptiles such as chameleons and is home to all the world's lemur species.

    5   **Indo-Burma**
        An area stretching from the eastern slopes of the Himalayas through Burma and Thailand to Indochina. This region hosts the world's highest freshwater turtle diversity (43 species), and a diverse array of mammals. Several new ungulate species, such as the saola and giant muntjac, were recently discovered here.

    6   **Caribbean**
        One of the highest concentrations of species per unit

area on Earth. Reptiles are particularly diverse (497 species are found here), 80 percent of which are found nowhere else. Non-fish vertebrates number 1518.

7 **Atlantic Forest region**
Once covering an area nearly three times the size of California, the Atlantic Forest has been reduced to about 7% of its original extent. It is most famous for 25 different kinds of primates, 20 of which are endemic. Among its best-known 'flagship species' are the critically endangered muriquis and lion tamarins.

8 **Philippines**
The most devastated of the hotspots, the forest cover has been reduced to 3% of its original extent. The Philippines is especially rich in endemic mammals and birds, such as the Philippine eagle.

9 **Cape Floristic Province**
This Mediterranean-type hotspot in southern Africa covers an area roughly the size of Ireland, and is now approximately 20% of its original extent. It is home to 8200 plant species, more than 5500 of which are endemic.

10 **Mesoamerica**
Forming a land bridge between two American continents, this hotspot features species representative of North and South America as well as its own unique biota. The spider and howler monkeys, Baird's tapir and unusual horned guan are 'flagship species'.

11 **Brazilian Cerrado**
A vast area of savanna and dry forest, the Cerrado is Brazil's new agricultural frontier and has been greatly altered by human activity in the past few decades. Home to 4400 endemic plants and several well-known mammal species, including the giant anteater, Brazilian tapir, and maned wolf.

12 **Southwest Australia**
A Mediterranean-type system, this hotspot is rich in endemic plants, reptiles, and marsupials including the numbat, the honey possum and quokka. It is also home to some of the world's tallest trees, e.g. the giant eucalyptus.

13 **Mountains of South-Central China**
An area of extreme topography, these mountains are home to several of the world's best-known mammals, including the giant panda, the red panda, and the golden monkey. This hotspot is largely unexplored and may hold many undiscovered species.

14 **Polynesia/Micronesia**
This hotspot comprises thousands of tiny islands scattered over the vast Pacific, from Fiji and Hawaii to Easter Island and is noteworthy for its land snails, birds, and reptiles. Hawaii has suffered some of the most severe extinctions in modern history, due in part to the introduction of non-native plant and animal species.

15 **New Caledonia**
One of the smallest hotspots yet it has the largest concentration of unique plants with five plant families found nowhere else on Earth. This hotspot also features many endemic birds, such as the kagu, a long-legged, flightless forest dweller representing an entire bird family.

16 **Choco-Darien Western Ecuador**
Some of the world's wettest rain forests are found here, and amphibians, plants and birds are particularly abundant. It has one of the highest levels of endemism of any hotspot with 210 endemic amphibian species of the 350 species found here.

17 **Guinean Forests of West Africa** (in error, this hotspot was not numbered on the map). With the highest mammalian diversity of any hotspot, these forests are home to the rare pygmy hippopotamus and many other striking species, including the western chimpanzee, Diana monkey and several forest duikers. The numbers of these endemic mammals have been severely reduced by large-scale logging and hunting.

18 **Western Ghats/Sri Lanka**
The Western Ghats mountain chain and adjacent island of Sri Lanka harbour high concentrations of endemic reptiles; of 259 reptile species, 161 are found nowhere else on Earth. This hotspot is also home to a number of 'flagship species', including the lion-tailed macaque.

19 **California Floristic Province**
Extending along the coast of California and into Oregon and northwestern Baja California, Mexico, this is one of five hotspots featuring a Mediterranean-type climate of hot, dry summers and cool, wet winters. It is especially rich in plants, with more than 4000 plant species, almost half of which are endemic.

20 **Succulent Karoo**
The only arid hotspot, the Succulent Karoo of southern Africa is renowned for unique succulent plants, as well as lizards and tortoises. in Namaqualand, in the southern part of this hotspot, a seasonal burst of bloom in September attracts many tourists.

21 **New Zealand**
This hotspot claims a number of world-famous endemic bird species, including kiwi (a nocturnal, flightless bird), takahe (a diurnal, flightless bird), and the critically endangered kakapo (a large, flightless parrot).

22 **Central Chile**
This hotspot features an arid region as well as a more typical Mediterranean-type zone. Best known for its incredible variety of plant species but also features unusual fauna, including one of the largest birds in the Americas, the Andean condor.

23 **Caucasus**
Situated between the Black Sea and the Caspian Sea, Caucasus habitats range from temperate forests to grasslands. A diversity of plants have been recorded here with some 6300 species, more than 1600 of which are endemic.

24 **Wallacea**
Named for the 19th century naturalist Alfred Russel Wallace, this hotspot comprises the large Indonesian island of Sulawesi, the Moluccas and many smaller islands. The area is particularly rich in endemic mammals and birds.

25 **Eastern Arc Mountains/ Coastal Forests of Tanzania and Kenya**
A chain of upland and coastal forests, this hotspot claims one of the densest concentrations of endemic plant and primate species in the world. It is home to African violets and 4000 other plant species, as well as the 1500 remaining Kirk's red colobus monkeys.

2. Student's own opinion as supported by an explanation. The major threats to biodiversity include: Population growth and resource consumption, over-hunting/commercial exploitation, illegal trading, habitat conversion and sprawl, establishment of exotic and invasive species, environmental degradation/pollution, and global warming.

---

## Tropical Deforestation (page 234)

1. (a) They enhance removal of carbon dioxide from the atmosphere (anti-greenhouse).
   (b) They maintain species diversity.
   (c) They have, as-yet-undiscovered, potentially useful species for medicines etc.

2. Tropical deforestation has three primary causes: (1) logging, (2) fires, and (3) road-building (associated with clearance for agriculture). Logging and fires destroy forest. Intrusion of roads into pristine forested areas allows the invasion of weed species, increases erosion, and prevents the reestablishment of forest species. Agriculture maintains cleared areas and prevents forest reestablishment. Continued agriculture on thin tropical soils precludes the easy reestablishment of forest once the agricultural land has been abandoned.

© 2007-2013 **BIOZONE** International
ISBN: 978-1-927173-60-2
Photocopying Prohibited

## The Impact of Alien Species (page 235)

1. Student's own choice. Examples could include Kudzu, a deliberately introduced climbing vine that has aggressively invaded the southern US, or the red imported fire ant which were accidently introduced into the US and now how displaced native ant populations in 14 states.

2. Alien species often have no natural controls within the new environment (e.g. no predators or natural competitors to keep their numbers in check). This means that they are able to rapidly reproduce and expand into the new area, often out-competing native or existing organisms as they do so.

## Endangered Species (page 236)

1. Students should give an answer appropriate to their species choice. Typically, recent extinctions are associated with hunting (e.g. dodo, passenger pigeon), severe habitat loss or fragmentation, introduction of alien species as competitors and/or predators (e.g. Stephen's Island wren).

2. (a) and (b) in any order:
   - Preserving biodiversity on Earth is ultimately of benefit to humans. Preserving biodiversity ensures the continued existence of species/foods/medicines/natural materials that could be of future use.
   - Other species have a right to exist alongside humans. As the main instigators of change on the planet, humans have a moral obligation to act as guardians of biodiversity for future generations.

3. (a) and (b) Students should give an answer appropriate to their own country or local region. Link into CITES and WWF through Biolinks for information. Typical reasons for decline (b) include: human pressure and habitat loss, degradation, or fragmentation, pressure from introduced predators or competitors, hunting/trade, introduction and spread of diseases as a result of contact with alien species.

## Conservation of African Elephants (page 237)

Note that the graph horizontal axis should include the label "Year".

1. In 1989, the African elephant was placed in Appendix 1 of CITES, which imposed a ban on trade in living or dead material from elephants.

2. (a) A limited legal trade in ivory has resulted from a policy of management and quota operation. Removal of the ban on ivory has allowed the rural communities of these countries to earn money from the controlled exploitation of their wildlife. (Advocates of this claim that it has dramatically increased the amount of land given over to wildlife, as the returns from wildlife exploitation have exceeded those from cattle).
   (b) Any two of:
   - Quota systems can be abused (and have been in the past, with illegal hunting continuing).
   - As returns from ivory increase there will also be pressure to extend the quota above what individual elephant populations can sustain.
   - As ivory is traded, there will be pressure to illegally bring in ivory (for trade) poached from vulnerable populations outside quota countries.

## In-situ Conservation (page 238)

1. *In situ* conservation uses ecosystem management and legislation to protect species in their natural habitat. By necessity, this involves both restoring the ecosystem and implementing laws to protect the species of interest. Methods include protecting and/or restoring the habitat, and protecting the endangered species from predators, hunting, and illegal trade (e.g. by CITES). Neither tool is effective in isolation; if the species are not protected, there is little point is restoring their habitat and if they are without habitat, there is little point in protecting them.

## Ex-situ Conservation (page 239)

1. *Ex-situ* conservation methods are often employed when species numbers become critically low or *in-situ* methods are not working. Features of *ex-situ* conservation focus on (1) removal of the endangered species from its natural habitat to a new location, usually a protected or controlled area and (2) captive breeding (or cultivation) involving a managed breeding (seed) register to maximize remaining genetic diversity.

2. Animal species do not all adapt equally well to captivity. Some species will not breed successfully in captivity or have very specific social requirements that might not be met by the conditions. Conversely, those individuals (and species) that adapt most successfully to breeding in captivity are inadvertently selected for survival in zoos, therefore their return to the wild might be compromised.

3. (a) Public education to promote species conservation and increase awareness.
   (b) To participate in global breeding programs to secure the viability of threatened species.
   (c) Zoos act as custodians for rare species; they cannot save a species from extinction if breeding is unsuccessful, but they can protect breeding individuals while other plans are implemented.

4. Gene banks and seedbanks provide a store of genetic diversity from wild stocks so that the genetic diversity is preserved in the advent of species loss or decline. Using modern reproductive technologies, gene banks can be used to boost the genetic diversity of inbred populations of endangered species. They also house genetic diversity from which to improve food crops and domestic livestock breeds. Wild species have characteristics (e.g. disease resistance, hardiness) that could be useful in agricultural production and it is important that this store of diversity is not lost.

5. *In situ* conservation involves a whole-ecosystem management approach to saving species in their own habitat. Methods include protecting or restoring the habitat, and protecting endangered species from losses (e.g. from introduced predators, illegal trade, and hunting). If whole-ecosystem restoration is successful it offers a good chance of species recovery, even for critically endangered species. It has the advantages of less disturbance to the species involved, it by-passes the need for captive breeding (which is unsuccessful for some species), and it offers a greater chance of long term success because habitat restoration goes hand in hand with species management. *Ex-situ* conservation methods are often employed when species numbers become critically low or in-situ methods are not working or are not feasible. *Ex-situ* conservation involves removing the endangered species from its natural habitat to a new location, where it can be monitored and protected more easily. *Ex-situ* methods rarely save species from extinction, and are often costly and labor intensive. Because the breeding stock is also limited, genetic diversity of the species may also become compromised.

## Conservation and Sustainability (page 241)

1. Conservation refers to the management of a resource so that it is maintained into the future. The resource may be a living or mineral system and may focus on the efficient use of a non-renewable resource or the managed use of a living system. Sustainability refers to the idea of using resources within the capacity of the environment and the eventual replacement of what has been used. It can be viewed as a subset of conservation, mainly focusing on living systems.

2. Conservation emphasizes system management; preservation aims to protect through isolation. Conservation manages resources actively and allows sustainable resource use, e.g. tourism in a forest. Preservation protects pristine areas without exploitation, e.g. National Parks with restricted access.

3. Societies build economies, which require resources in ever-increasing amounts as they grow. The rate of societal and economic growth may be greater than the rate of resource replenishment and, if so, the environment will suffer. In order for the environment to flourish, there must be a balance

met between the growth of society, economic gains, and the environmental systems and resources affected or used.

### Saving the Black Robin (page 242)
1. Black robin populations appear to have been low for a long period of time, but deteriorating habitat put them in critical danger of extinction. The robin was too small to fly to other habitats and the number of birds was decreasing rapidly.

2. The main method to increase the population size was cross fostering. Eggs were taken from the black robin nest and placed in the nest of other birds (first the Chatham Island warbler, then later the Chatham Island tomtit). The loss of eggs induced the black robin to lay again so that multiple clutches of eggs were produced each season.

3. (a) Old Blue was effectively the fittest of the black robins, producing offspring that survived to reproduce so that all black robins are related to her.
   (b) There is a risk of inbreeding causing genetic problems as the genome of each black robin is almost identical to any other black robin.

### The Sixth Extinction (page 243)
1. Estimating the number of species on Earth is difficult because so many parts of the Earth remain undocumented, including most of the sea and much of the rainforests. There is also so many species that may or may not be separate species (depending on how species are defined). The difficulty in estimating the number of species on Earth makes estimating their extinction rate even harder. Species that are completely unknown to science may be becoming extinct every day, thus dramatically increasing the extinction rate from current estimates.

2. (a) Mammals: 17.4 extinction per century. (5487/1,000,000) x 100 = 0.5487. 17.4/0.5487 = 31.7 times greater than the background rate.
   (b) Reptiles: 4.4 extinctions per century. (10,000/1,000,000) x 100 = 1. 22/1 = 22 times greater than the background rate.
   (c) Amphibians: 7.8 extinctions per century. (6700/1,000,000) x 100 = 0.67. 7.8/0.67 = 11.6 times greater than the background rate.
   (d) 22.8 extinctions per century. (300,000/1,000,000) x 100 = 30. 22.8/30 = 0.76 of the background rate.* *This rate is almost certainly an underestimate due to the number of unknown plants (especially those in rainforest which are rapidly being destroyed).

3. The Sixth Extinction is the apparently human-induced loss of much of the Earth's biodiversity.

4. Humans may not directly or deliberately intend to cause the extinction of an organism, but often when they enter a new environment they introduce new organisms (including diseases) and modify the environment. The new organisms may be parasites, predators, or competitors that reduce the fitness of the resident organisms to the point that the populations are unsustainable. Modifying the landscape, e.g. by deforestation or damming rivers, may reduce usable habitat for resident organisms and introduce pollutants, which affect survival and reproduction.

5. Humans tend to have a natural affinity to what people call "cute and cuddly". These animals tend to keep a baby-like or soft toy-like look throughout their lives. Animals such as meerkats, pandas, seals, and big cats all appear this way. Animals such as spiders and snakes tend to produce an aversion reflex in humans. As a result of this we know less about the unattractive looking creatures than the attractive ones. This affects their survival in that what is not known can not be used in conservation efforts., i.e. if an animal is not studied it is impossible to tell if it is endangered or not.

6. Examples include:
   Quagga 1867, hunting
   Pyrenea ibex, 2000, hunting
   Bubal hartebeest, 1923, hunting
   Tecopa pupfish, 1981, habitat loss.
   Baiji river dolphin (Yangtze River dolphin), 2006, hunting, habitat loss, and pollution of habitat.
   There are, unfortunately, many more.

### KEY TERMS: Mix and Match (page 25)
Biodiversity (B), Carbon credit (A), Deforestation (G), Endangered species (J), Ex-situ conservation (I), Extinction (R), Global warming (Q), Greenhouse effect (N), Greenhouse gas (P), Habitat Restoration (C), Indicator organism (S), In-situ conservation (D), National Park (K), Ocean acidification (E), Ozone depletion (O) Seedbank (M), Stratospheric ozone (H), Threatened species (L), Ultraviolet radiation (F)